James Hamilton

Observations on the Seats and Causes of Diseases

illustrated by the dissections of the late Professor Morgagni of Padua - Vol. 1

James Hamilton

Observations on the Seats and Causes of Diseases
illustrated by the dissections of the late Professor Morgagni of Padua - Vol. 1

ISBN/EAN: 9783337255350

Printed in Europe, USA, Canada, Australia, Japan

Cover: Foto ©berggeist007 / pixelio.de

More available books at **www.hansebooks.com**

OBSERVATIONS

ON THE

SEATS AND CAUSES

OF

DISEASES,

&c.

OBSERVATIONS

ON THE

SEATS AND CAUSES

OF

DISEASES:

ILLUSTRATED BY THE

DISSECTIONS

OF THE LATE

PROFESSOR MORGAGNI OF *PADUA.*

BY

JAMES HAMILTON, Junior, M. D.
FELLOW OF THE ROYAL COLLEGE OF PHYSICIANS OF EDINBURGH.

VOL. I.

EDINBURGH:
PRINTED FOR P. HILL, EDINBURGH; AND
G. G. & J. ROBINSONS, LONDON.

1795.

BARTHOLOMEW PARR, M.D.

FELLOW OF THE ROYAL SOCIETY OF EDINBURGH, PHYSICIAN
TO THE DEVON AND EXETER INFIRMARY, ETC.

SIR,

I wish I could offer a better proof of attention to the frequent and valuable advice, with which you honoured me at a very early period of my life, than what the following Work contains.

As it is, I have ventured to lay it at your feet; since it affords me an opportunity of assuring you, that neither the distance at which we are placed, nor the avocations in which we are severally engaged, can ever erase from my mind the sense I entertain of the favours you have conferred upon,

SIR,

Your much obliged

Humble Servant,

JAS. HAMILTON, JUNR.

Edinburgh, Castlehill,
Jan. 20. 1795.

PREFACE.

A view of the morbid appearances in dead bodies, enables the medical practitioner to afcertain the caufes of fome difeafes, and the confequences of others; and, therefore, can be ufeful only when connected with the fymptoms which preceded death, and when applied to explain fuch fymptoms.

On this principle, THEOPHILUS BONETUS publifhed the Sepulchretum Anatomicum. But that work, even improved as it was by Mangetus, although in many refpects valuable, being neceffarily a rude fketch only, the celebrated Profeffor MORGAGNI of Padua undertook to render it more perfect, by the publication of Seventy Letters on the Seats and Caufes of Difeafes.

b   The

The great abilities, and extraordinary in-
dufiry, difplayed by MORGAGNI in thefe
letters, defervedly entitle them to a much
higher panegyric than it would be proper
for the Editor of the following work to be-
ftow.

So extenfively, however, has medical
knowledge been diffufed fince the publi-
tion of MORGAGNI's writings, although no
more than a period of about thirty years
has elapfed, that they are now deemed
chiefly valuable on account of the facts
which they contain. As, therefore, readers
in general find they have to wade through
a prodigious quantity of uninterefting matter, before they can arrive at what they regard as ufeful; thefe writings are at prefent
feldom looked into, being only occafionally
confulted as a dictionary.

The Editor of the following pages was induced, by thefe confiderations, to believe,
that, without deviating from the principles
of

of MORGAGNI's plan, such alterations might be made on the original work, as should contribute to render the many important facts with which it abounds extensively known, and consequently generally interesting.

No sooner had he formed the design of making an attempt of this kind, than he laid it aside, on Dr. Baillie's proposed publication on the Morbid Anatomy of the Human Body being announced. With great pleasure he yielded the task to one so much better qualified to undertake it. But when that work appeared, he found that Dr. Baillie's views were very different from his own; and, flattering himself that the labours of both might be severally useful, he again resumed his intention.

The principles, by which he resolved to be directed in the execution of his project, were these—To retain only the facts witnessed by MORGAGNI, or his preceptor VAL-

SALVA,

SALVA, or that feem eftablifhed on unequivocal authority—to new arrange the whole
—to prefix to each collection of cafes, a view
of the general fymptoms, and feat of the
difeafe—and, to add, obfervations on the
caufes, and remarks on the hiftories, detailed.

The firft part of his tafk was fufficiently
eafy. Not fo the fecond; for he felt it very
difficult to fix upon a fuitable arrangement;
and befides, fo many cafes were in the original claffed under erroneous titles, that it required much time to affign to each its proper rank *.

As he has long thought, in common with
many others, that, in nofology, difeafes ought
to be fo claffed, that thofe which are fimilar
may be collected together, he adopted the
arrangement

---

* Within a parenthefis added to each cafe, the number of the letter, and of the article under which it is placed in the original, is marked.

arrangement of Macbride; dividing difeafes into Univerfal, Local, Sexual, and Infantile. This method feemed more analogous to that of MORGAGNI, and better adapted to his facts, than any other.

In the tranflation of the facts, the Editor has taken fome liberties with the original, which perhaps may require an apology. He has omitted many of the little attending circumftances mentioned by MORGAGNI; fuch as the dates of the cafes, the names of the patients, &c.; and he has uniformly tranflated the meaning, rather than the peculiar expreffions. As to the former, he thought the dates, &c. little interefting to the general reader; fince they tend only to eftablifh the authenticity of the cafes, which feems quite unneceffary. With refpect to the latter, he has always confidered it to be the duty of the tranflator of a work, on any art or fcience, to communicate the meaning of the original in the moft accurate ftyle of

which

which language is fufceptible. This rule feemed peculiarly applicable to the work of MORGAGNI; for the ftyle of the whole is exceedingly uncouth.

But he has ventured ftill farther. In the original, many hiftories are only partially detailed; detached parts of them being inferted under different heads, and fometimes even in other works of the author. Thefe he has brought together; his great aim having been to exhibit in a perfpicuous manner every fact detailed.

In his tafk as a tranflator, the Editor thinks it incumbent on him to acknowledge the great affiftance which he derived from the tranflation by Dr. Alexander. It leffened his labour, by ferving as a fketch.

The hiftory of the general fymptoms of each difeafe, prefixed to the refpective cafes, has been compiled with care from fuch fources as feemed moft authentic; and to thefe proper references are made.

The

The obfervations on the caufes of difeafes, comprehend the moft modern opinions; and with thefe the ideas of MORGAGNI are occafionally incorporated. In this part of the work the Editor has endeavoured to fay much in very few words; and on that account has avoided all minute reafoning, and has contented himfelf with ftating only the chief arguments on every fubject. Where he has diffented from others, he has expreffed himfelf concifely, and he hopes with becoming diffidence. In treating of the caufes of difeafes, he has always wifhed rather to difcover the deliberation of a found judgement, than to difplay the fportings of a lively imagination. Though by the latter, the ignorant and unwary may be dazzled into admiration; it is upon the former alóne, that the experienced and cautious will be inclined to depend.

The remarks he has added to the cafes, being confined ftrictly to the explanation of

the

the refpective difeafes, are neceffarily very
fhort in the volume now offered. In the
fubfequent part of the work, references to
many of the cafes are occafionally made,
and additional obfervations upon them are
introduced.

The Editor hopes to complete that part of
the work, allotted to Univerfal Difeafes,
within another volume; or, fhould the materials be found incompreffible into that fize,
he engages that it fhall not exceed two vo-
lumes.

It is meant that the Local, Sexual, and
Infantile Difeafes, fhall be continued in a
feparate publication, conducted on the fame
plan.

CONTENTS.

c   § 2. Tabes

Caufes

PRELIMINARY OBSERVATIONS.

As nothing can be more prejudicial to the fludy of any art or fcience than ambiguity in the language employed, it is neceffary to premife, as an introduction to the following pages, the explanation of what is meant by the terms expreffive of the different caufes of difeafe.

The word Caufe is, throughout the whole work, ufed according to the common acceptation of the term: MORGAGNI was no metaphyfician ; and no more is the Editor of thefe pages.

The Caufes of difeafe are divided into the *Predifponent*, the *Exciting*, and the *Proximate*.
The Predifponent caufe, is that circumftance, or combination of circumftances, by which the body is rendered fufceptible of any particular difeafe.
The Exciting caufe, is that circumftance, on the application of which to the body difeafe follows.
The Proximate caufe, is that circumftance, or combination of circumftances, from which the
symptoms

symptoms of the difeafe arife. It is therefore the effect of the application of the exciting caufe.

Thefe definitions may be illuftrated by a familiar example.

Perfons who have a delicate habit and florid complexion, and at the fame time a long flender neck and narrow cheft, are much fubject to a difcharge of blood from the lungs. That particular conformation of the fyftem is therefore regarded as the predifponent caufe of the difeafe alluded to.

When in fuch perfons violent paffions of the mind are excited, or irregular action of the lungs takes place, or any accuftomed evacuation is fuddenly ftopt, or the blood is circulated through the veffels with undue force, or when they indulge in a larger than ufual proportion of food, and at the fame time ufe lefs exercife, and a difcharge of blood from the lungs enfues—One or more of thefe circumftances is to be confidered as the exciting caufe or caufes of the difeafe.

The effect of the application of thefe exciting caufes to the fyftem of a perfon of that defcription, is the laceration or divifion of one or more bloodveffels within the lungs; and hence fuch laceration is deemed the proximate caufe of the difeafe.

The predifponent and proximate caufe, therefore, muft exift in the fyftem of the perfon affected;

whereas

whereas the exciting may be fome external circumftance.

Although, in general, the application of the exciting caufe produces no effect, unlefs the predifponent previoufly exift; yet there are many exceptions to this rule. But the proximate caufe can never take place, without being preceded by an exciting one.

A knowledge of the predifponent and exciting caufes of morbid affections enables the phyfician to prevent difeafes; and that of the proximate caufe has been imagined neceffary to direct him in the cure.

The predifponent and exciting caufes are in many inftances eafily afcertained, and for a good reafon, as they are difcovered principally by obfervation. The latter are known by that means alone.

The proximate caufe however cannot be afcertained, without an intimate acquaintance with the ftructure and functions of the human body.—But as fuch knowledge is yet in a very imperfect ftate, the proximate caufe of difeafes is ftill involved in fo much obfcurity, that it is difcovered only in thofe diforders which are feated in a fingle organ, and in fome particular part of the ftructure of that organ.

UNIVERSAL DISEASES.

CHAPTER I.

FEVERS.

SECT. I. *CONTINUED FEVER.*

§ 1. CONTINUED INFLAMMATORY FEVER.

ACCORDING to Dr. Cullen's definition, Continued Inflammatory Fever, or what he calls Synocha, is that where, along with the general character of pyrexiæ, the heat of the body is very much increased; the pulse is frequent, full, and hard; the urine is red; the functions of the brain are but little deranged; and there is no primary local difeafe.

As this fever occurs very feldom, if ever, in this part of Great Britain, the defcription of its fymp-

toms is neceſſarily taken from foreign authors, or
from thoſe who have copied from them *.

This fever is uſhered in by ſlight rigors †, without
any previous languor or ſigns of debility. The
tongue is white and dry, attended with a bad taſte
in the mouth and thirſt. The ſenſe of ſmelling is impaired, and nauſea is felt. A hot fit very ſoon ſuc-
ceeds; the pulſe, which before had been ſmall and
depreſſed, becomes full, ſtrong, and hard; and the
heat of the body is ſo much increaſed, that it is equal
to a degree not under 106 of Farenheit's ſcale; and
ſometimes exceeds that degree conſiderably. The
face is much fluſhed; the eyes are inflamed, and
impatient of bearing light. Headach, and ſometimes pain of the back, take place. The tongue is
black and parched, and the thirſt continues exceſſive. The urine is high coloured, and the belly is
coſtive. The breathing is generally laborious;
and is ſometimes attended with a ſhort cough.

In favourable caſes, a ſweat breaks out after
the hot fit; and the diſeaſe is thereby carried off
within twenty four hours from its commencement.
In other caſes, the ſkin continues hot and dry;
laborious

---

* Vide Junker Conſpect. Therap. Special. pag. 484. Lommii
Medicinal. Obſervat. p. 7. Liddelii Opera, p. 78. Eller Obſervat.
de cognoſcend. et curand. Morbis, p. 62. Home Princip. Med.
p. 76. Meza Compend. Med. Pract. § 128, &c. Oſterdyk Præ-
cept. Med. Pract. p. 38.

† Junker and Lommius allege that it is not preceded by rigors,
but invades ſuddenly with great heat.

laborious breathing and headach remain unabated; reftleſsneſs or very diſturbed ſleep ſupervenes, attended with tinnitus aurium, and the appearance of motes floating before the eyes; and theſe ſymptoms are ſucceeded by delirium. In ſuch caſes, it ſometimes happens that a double pulſation in the arteries can be perceived. Epiſtaxis, preceded by increaſed pain in the head and tinnitus aurium, violent throbbing at the temples and itching of the noſe, ſometimes occurs about the fourth or ſeventh day, and proves critical. The criſis, in other caſes, is by urine, vomiting, ſweating, or diarrhœa. If, however, yellowneſs of the ſkin appear before the ſeventh day, or if delirium, coma, or convulſions take place within the ſame period, the diſeaſe terminates fatally.

During the courſe of this fever, the pulſe continues quick, full, and hard, until towards the criſis or termination of the diſeaſe. In favourable caſes, it ſeldom exceeds one hundred and twenty or one hundred and thirty pulſations in a minute. The ſtate of the tongue varies from being white and dry to a black colour, and to ſo great a degree of dryneſs that it is ſometimes cracked. The urine, at firſt high coloured and in ſmall quantity, becomes either natural, or in large quantity with a copious ſediment, or bloody, or black and fetid, according to the nature of the event of the diſeaſe. The belly either continues obſtinately coſtive, or

A 2

diarrhœa

diarrhœa occurs.   Night exacerbations are pretty
diftinctly marked.   Great anxiety and defponden-
cy generally attend.   Ulcers and iffues, it is faid,
during the courfe of this fever, are dried up; and
affume an inflamed appearance.   Blood drawn
during the difeafe is fizy.

This fever fometimes terminates in inflammation of the lungs.   Perfons under forty years of
age are the moft ordinary fubjects of this difeafe;
but, as the fucceeding cafes fhow, perfons much
beyond that age are not exempted from it.

The appearances on diffection, where inflammatory fever terminates fatally, generally exhibit
effufions within the cranium, a determination of
blood to the veffels of the head, and fometimes
even fuppuration within the brain.

From the fymptoms of the difeafe, and from the
appearances on diffection, it is probable that the
feat of inflammatory fever is the fanguiferous fyftem.

CASES OF CONTINUED INFLAMMATORY FEVER.

CASE I.   (x. 17.)

A YOUNG man, aged twenty five years, by trade
a wool-comber, affected with continued fever, became fo delirious that it was neceffary to bind him.
                                    The

The delirium having remitted, he was brought into the hofpital of Padua. Immediately on his admiffion, convulfive motions of his fuperior extremities, and fubfultus tendinum at the wrifts, were obferved. Venefeftion having been ordered, the blood exhibited no inflammatory cruft; but its fubftance was denfe and compact. He became quite comatofe; and, having fpoken none for the laft three days, died.

Appearances on Diffection.

THORAX. Nothing remarkable was obferved, in this cavity, except that one of the lobes of the lungs was hard. In feparating the fifth dorfal vertebra from the fixth, a confiderable quantity of fluid flowed out from the fpinal tube. When the veffels in the neck were divided, much blood of a black colour was difcharged.

HEAD. When the cranium was fawed through, a fmall quantity of fluid, fimilar to that in the fpinal tube, iffued out. The upper part of the cranium and the portion of the dura mater under it being removed, the veffels of the pia mater, on the pofterior part of the left hemifphere of the brain were obferved to be diftended with black-coloured blood. Under the pia mater a kind of jelly was feen in feveral places, with air bubbles intermixed, although no bad fmell was perceived in any part of the body; and it was not probable that, at that feafon of the year, a body fhould become putrid

trid within lefs than three days after death. The
fubftance of the brain, when cut into, was found
to be very hard ; bloody points appeared throughout its medullary fubftance, which forming immediately into large drops of blood, afforded ample
proof of the prefence of much fluid blood. The
lateral ventricles contained fcarcely any ferous
fluid ; they were fhorter than ufual. The plexus
choroides were of a black red colour. The pineal
gland was of a rofy colour; its anterior and pofterior furfaces were not depreffed, as they ufually
are, but rather turgid and full. It was very hard,
and when cut into was found to contain feveral
fmall calculi as it were. One of thefe refembled
a millet feed, both in magnitude and form ; but
was of the hardnefs of bone, and feemed alfo, from
the fmell which it emitted when applied to the
flame, to be of an offeous nature. The fpinal
marrow was accurately examined from its origin
down to the fifth dorfal vertebra. All the veffels
of the continuation of the pia mater, efpecially
thofe on the pofterior furface, were fo much diftended with blood that they refembled veffels
which had been injected with red wax. The fanguiferous veffels which accompany the fpinal
nerves, efpecially fome of thefe nerves, were alfo
obferved to be diftended with blood.

CASE

Case II. (1. 12.)

A young woman, the wife of an indigent man, and the daughter of a woman subject to epilepsy, in consequence of being overheated after a journey, (in the month of February), was affected with a violent pain in the head and an ardent fever. These symptoms continuing, she died within the space of three or four days, having had no delirium, but having been often reservedly silent. When affected with the disease, she gave suck, and at the same time had the catamenia. For these reasons blood-letting had been delayed so long, that when, from the symptoms becoming worse, although the pulsation and strength of the arteries continued firm, half a pound of blood was drawn from the foot, it so happened that she immediately expired. The blood instantly coagulated very strongly.

Appearances on Dissection.

Head. The inside of the skull was of a brown red colour. The external surface of the pia mater on the upper part of the brain was covered with a yellowish fluid, in no great quantity, but spread equally over it. Its consistence was somewhat thick; and, although perfectly inodorous, it had altogether such an appearance that it seemed to be really purulent matter. The cerebrum was discoloured; but no mark of disorder could be perceived in the meninges

ges or brain, nor could any traces of the origin of
the pus be difcovered.

CASE III. (LXII. 5.)

A SCAFFENGER, apparently aged about fifty
years, of a robuft habit, and of a healthy appearance, but rather plethoric, and addicled to
drunkennefs, was employed, along with his fervants, in cleaning out the jakes of an hofpital, at an
unfeafonable hour of the night, a time which is generally chofen for fuch bufinefs. As they were neceffarily going to and fro, the mafter himfelf being
left at one time alone, imagined that he faw a
fpectre clothed in white, and was immediately affected with univerfal tremor, while at the fame
time his mouth was diftorted. In this ftate he was
found by his fervants, who carried him inftantly
to bed. Antifpafmodics and cordials were immediately given; and the tremors having remitted,
and the pulfe become ftrong, half a pound of blood
was drawn from his arm that night. In the morning, as much blood was taken from the other
arm; for the tremors had remitted completely,
and the pulfe had become more full and febrile.
On the following day alfo a vein was opened in the
foot, as fome alleviation of the fymptoms, though
for a fhort fpace of time, was perceived after each
bleeding. The blood, efpecially in the firft bleed-

2   ings,

ing, came out in a frothy ftate, of a very black colour; the craffamentum was rather hard, and the ferum in fmall quantity. The fever however continued; and, inftead of the tremulous convulfive motions with which he had at firft been affected, his whole body, from time to time, was agitated with violent tonic convulfions. He could not fpeak intelligibly, nor had he done fo from the time that he had related to his fervants what had happened to him. It was evident, however, that he knew the perfons who were about him, and could diftinguifh them from one another. When he was able, he fignified, by means of geftures, that he was affected with a very fevere pain in the head. In confequence of thefe fymptoms, to alleviate which feveral external and internal remedies, befides thofe mentioned, were in vain employed, he died, within fix or feven days from the commencement of the difeafe.

Appearances on Diffection.

EXTERNALLY. The penis and fcrotum were of a black colour, but the fkin only was affected; the fingers were very rigid, but not the arms.

ABDOMEN. When the omentum, which had very little fat, was removed, the colon, in its whole extent diftended with air, though not immoderately, was obferved to follow fuch a direction, that, after having afcended to the liver, it defcended from that to two or three inches below the navel on the right fide; and having returned from that to its ordinary fituation, which it retained as ufual in going acrofs under the ftomach, lying in an oblique direction in the left fide of the left hypochondriac region, and in a ftraight direction over the whole anterior furface of the left kidney, it again returned into the fame hypochondre, and defcending from thence, and entering the pelvis, it terminated on the rectum without any previous flexure. The fmall inteftines, except fome tracts of them, efpecially a confiderable portion of the ileum which lay very low in the pelvis, were diftended with air. A yellow colour with which they were internally tinged, was feen through their coats; for the bile, with which the gall-bladder was almoft filled, had by exudation made the contiguous inteftines externally yellow, and by flowing into them had rendered their internal furfaces of the fame colour. The liver and fpleen were of a leaden colour; but although at the edge of the former vifcus that colour was deeper, it did not in either extend beyond the furface. The fpleen was of a moderate fize; the liver was large; but both were found.

THORAX. The lungs fcarcely adhered to the pleura: In the few places where they did, it was at the pofterior part. They were rather turgid; there appeared, in fome places, pretty large vefications fomewhat raifed above the other parts of

.the

the furface ; a kind of veficles, as it were, feemed evidently to be included in thefe. There was no intermediate lobe on the right fide, but the appearances were the fame on it as in the left. The pericardium did not contain a fingle drop of fluid ; its internal furface was ftill moift, fo that it did not adhere to the heart, except very flightly in fome places. In the heart there was almoft no blood. This might perhaps happen from the blood having flowed out, in confequence of the large veffels below the diaphragm having been cut through, efpecially as the blood had been found fluid in feveral parts of the body. Two polypous concretions, however, were found in the heart. One of thefe extended from the right auricle into the vena cava fuperior; the other, which was fomewhat more remarkable, being round, and thicker than one's little finger, alfo extended from the right ventricle into the pulmonary artery.

HEAD. Nothing remarkable appeared when the cranium and dura mater were cut through. But the veffels of the pia mater were fo full of blood, that the fmalleft trunks feemed as if filled by injection. The veffels of the ventricles, and within the medullary fubftance of the brain, were alfo diftended ; and when the beginning of the fpinal marrow, which had been taken out together with the medulla oblongata, was gently compreffed, blood was obferved to iffue out, not only from the

B 2 fiffure

fiffure in its medullary fubftance, but alfo from the fection of the cineritious fubftance which was neareft the fiffure. A confiderable quantity of limpid fluid was obferved in both lateral ventricles; but the plexus choroides were red. No veficles were perceived on them as there ufually are, but feveral very fmall red particles, which were folid fo as to feem glandular, were obferved. The cerebrum and cerebellum were of the natural firmnefs. The fornix was flabby, as was alfo the internal furface of the trunk and crura of the medulla oblongata. A portion of the furface of the anterior lobes of the cerebrum, at the middle of the higheft part where they are contiguous to each other, was fo formed, that the one lobe was received into a hollow of the other. The remaining part of their furface was convoluted in the ordinary manner.

CASE IV. (VII. 17.)

AN old woman was affected with a flight fever, which her phyfician hoped to have removed by the ufe of Peruvian bark. It did not, however, yield to this treatment; but, on the contrary, having become an acute fever, attended with flight wandering of the mind, fhe died.

Appearances on Diffection.

ABDOMEN. Two ureters proceeded from the
right

right kidney; the superior one was small, arising from a very simple pelvis; the inferior was thicker, as it proceeded from a pelvis which was rendered larger and more prominent, in consequence of many tubuli terminating in it. The progress and insertion of each ureter was as distinct as their origin; for the orifices of both opened at the distance of about a finger's breadth from the other into the bladder, in the same oblique direction as usual, in such a manner that the one was above the other. In the superior and posterior part of the fundus uteri an excrescence of a round form, and externally of a bloody colour, was found extending from the right side towards the left. Nearly a third part of its circumference, at the inferior and left side of it, was separated from the uterus so that it could be raised by the probe; the remaining portion was intimately connected with the uterus, and indeed seemed to be composed of the same substance; but, when cut into, it was found to be throughout of a paler colour, and harder and more compact. This was certainly the beginning of a scirrhus, or perhaps of an ocult cancer in the very mildest state. It was flat, smooth, and so small that it could be covered with the first joint of one's thumb when extended. From the structure of the surface contiguous to the cervix uteri, and the appearance of the hymen, the edges of which were not broad, but were entire, it was

evident

evident that the woman had had very little if any commerce with man.

THORAX. In the heart a kind of membrane, perforated like a fieve, or of a ftructure refembling net-work, was found to occupy the place of the valve of the coronary vein. Incipient offifications, of a white colour, were feen in the internal furface of the aorta, a little above the femilunar valves, and at that part placed at the lumbar vertebrae.

HEAD. The brain was accurately examined; but nothing deferving notice appeared, except that the veffels of the pia mater were diftended with blood. That membrane, it may alfo be obferved, was eafily feparated from the brain in every place; and confequently, although little fluid was feen, it was evident that there muft have been fome.

C A S E V. (XI. 22.)

A WOMAN who had formerly had an apoplectic paroxyfm, was again affected with the fame difcafe, and afterwards remained in a ftupid and femiparalytic ftate. Within one or two months from that period fhe became affected with fever, which was very violent; as the ftate of the pulfe and the great thirft plainly indicated. Of this fhe died in the hofpital of Bologna.

Appearances

Appearances on Diffection.

THORAX. A polypous concretion was found in each fide of the heart. That on the left had feveral branches; its trunk was more firm than that in the other fide: externally it appeared furrounded with a tendinous fubftance; and internally its fubftance refembled that of firm compact flefh.

HEAD. The veffels of the cerebrum were in fome degree turgid with black blood. The fubftance of the brain was fo foft, that when in fome places the dura and pia mater were drawn off, the cortical fubftance followed. A fmall quantity of watery fluid was found in the third ventricle.

CASE VI. (VII. 7.)

A MAN, aged thirty-five years, was affected with a violent fever. He became delirious; his eyes gliftened; and his pulfe was quick and ftrong. At laft he died.

Appearances on Diffection.

HEAD. The blood veffels of the brain were turgid from the contained blood. In the ventricles a fmall quantity of ferous fluid was found. The brain was in a found ftate.

The blood in every part of this body was fluid, except in the heart, in which fome polypous concretions were obferved.

I   CASE

C A S E VII. (IX. 12.)

A MAN, by trade a, cook, formerly fubject to
difeafes of the urinary paffages, affected violently
with continued fever, was received into the hofpital at Bologna. Having been bled, the blood
became fo much coagulated that it adhered ftrongly to the fides of the glafs veffel in which it
was kept; and all the ferum, which was in fmall
quantity and bloody, was forced out on its top.
He continued to grow worfe, efpecially at night.
About the twelfth day of the fever epileptic paroxyfms fupervened, and he died.

Appearances on Diffection.

ABDOMEN. One of the kidneys was round: it
refembled fomewhat a cancer, and contained calculi. The other kidney was twice as large as u-
fual, probably from its performing the office of
both.

THORAX. The pleura was inflamed. The heart
and large veffels were diftended with very black
blood; which was very fluid, and ftill warm, although ten hours after death.

HEAD. All the veffels which creep over the
furface of the brain appeared very red and exceedingly turgid. A fmall quantity of watery
fluid, clear as lymph, was found in the ventricles.

CASE

CASE VIII. (XLIX. 10.)

A WOMAN, aged twenty-five years, of a bilious temperament, affected with difficulty of breathing, was admitted into the hofpital of Bologna. Along with the difficulty of breathing, fhe complained of pain in the left fide of the thorax; within which, during refpiration, a found like that of matter was diftinguifhed. Her pulfe was quick, though foft. On the fifth day from the date of her admiffion, the jaundice fupervened; and after having continued till the eighth day, it difappeared. At this time, from the obftinacy of the fever, blood letting, which had been employed on the firft days of the complaint, was again had recourfe to. Although the fever was not then fo violent as to threaten death, fhe died fuddenly.

Appearances on Diffection.

ABDOMEN. Every thing within the abdominal cavity appeared found, except that it contained half a pound of watery fluid; a circumftance by no means uncommon.

THORAX. The right lobe of the lungs, which adhered to the ribs at its upper part, was at its lower part inflamed; and when its fubftance was there cut into, a little ferous fluid was difcharged. The left lobe was totally unconnected with the pleura, and was perfectly found. A polypous

VOL. I.   C   concretion

concretion extended from the right ventricle of the heart (in the mufcular fibres of which its bafis was placed) into the vena cava. It was of a firm ftructure; and its colour was at one part pale, and at another red.

C A S E IX. (XXXVIII. 22.)

A WOMAN, about thirty years of age, after long continued pains in the joints, became affected with a very copious, moift, fcabby eruption. In order to repel this, fhe ufed fome kind of ointment, by the advice of an empyric. By this means, the eruption was indeed dried up within a fhort time; but an acute fever, attended with great heat, with thirft, and with moft excruciating pains in the head, was the confequence. To thefe fymptoms, delirium, great difficulty of breathing, flight fwelling of the whole body, and confiderable fwelling of the abdomen, together with much reftlefsnefs, fupervened. On the fixth day from the time that the fever had forced her to keep bed, fhe died.

Appearances on Diffection.

ABDOMEN. When the belly, which was fwelled and very tenfe, was opened, inftead of water, the inteftines and ftomach burft out. Thefe contained nothing but air; with which they were fo much diftended that the ftomach filled more than half

of

of the abdominal cavity. In that cavity, about a pound or more of limpid ferum was found effufed; which when expofed to the fire, feemed at firft to be flightly coagulated, but afterwards, like the fluid of the pericardium, was entirely evaporated, leaving only a kind of yellow pellicle at the bottom of the veffel.

THORAX. The lungs adhered to the pleura by fuch a number of membranous fubftances refembling a gelatinous body, that it appeared they could not have been dilated fo freely as ufual. When thefe membranous fubftances were cut into, a pellucid fluid was difcharged. The heart at the right fide was connected with the pericardium by fome membranous fibres. Its ventricles contained fome fluid blood; and in the right ventricle, the beginning of a fmall polypous concretion was obferved.

In the diffection of the body, it was found that, when the fkin and flefh were cut into, no watery fluid was difcharged; from whence it was evident that the univerfal fwelling, mentioned in the hiftory of the cafe, did not proceed either from œdema or from anafarca. A circumftance which was alfo confirmed by the feet not having pitted upon preffure.

CASE

C A S E X. (LV. 10.)

A WOMAN, aged forty years, addicted to the
ufe of tobacco and wine, who had been married
to a robuft man, by trade a porter, but had never
had any children, having been affected with a
fcabby eruption, had repeatedly drank fulphur
mixed in wine, with a view to get rid of it. Having at laft taken a larger than ufual quantity of
the fame medicine, fhe began immediately to feel
indifpofed, and vomited repeatedly. She was foon
after brought into the hofpital at Padua, and was
then feverifh; had a hard and fmall pulfe, and
complained of great difficulty of breathing. Venefection was performed, and frefh drawn oil of
almonds was given. During the fucceeding day,
as the difficulty of breathing was ftill more confiderable, blood was again taken away; which, like
that formerly drawn, had a firm compact craffamentum, but was not covered with any cruft. Two
cups-full alfo of milk were given when the caufe
of the difeafe was learned. The difficulty of
breathing, however, having increafed, evident convulfions of the extremities having fupervened,
and the pulfe having become more languid, fhe
died about the fourth day from the beginning of
the difeafe.

Appearances

Appearances on Diffection.

EXTERNALLY. The body was in good condition, except that the fkin was here and there deformed with a flight fcabby eruption.

ABDOMEN. The belly was fwelled; not from too great a proportion of fat, for that was both in proper condition and in proper quantity; nor from the extravafation of watery fluid, for, although there was fome effufed fluid, it was entirely confined to the pelvis; but from the ftomach, the fmall inteftines, and confiderable tracts of the colon, being diftended with air. The colon, in other parts, either preferved its natural width, or was very much contracted, as was obferved in the left fide near the ftomach. The omentum covered none of the inteftines, as it was forced upwards or retracted; both it, and that fuperior portion of the mefocolon which fupports the tranfverfe arch of the colon, were rigid, and were here and there, efpecially on the pofterior part, marked with red fpots. On the external furface of the ftomach, the blood veffels were fomewhat turgid. On the internal furface, at the fundus, near the antrum pylori, there was an area of a circular form, the diameter of which was about four fingers breadth. It was diftinguifhed from the remaining furface of the ftomach by being lefs fmooth and lefs fhining, but more white, and being furnifhed with blood-veffels, which were black, as if from injection; whereas the remainder

mainder was more fmooth, more fhining, and lefs
white in colour; and exhibited almoft no veffels,
at leaft none fo diftinct or black. The internal
coat of the ftomach appeared eroded throughout the whole extent of the area. Except in that
part, no mark of erofion or inflammation could
be traced, either in the ftomach, or in the adjoining part of the œfophagus or inteftinal canal.
The ftomach was larger than ordinary; it had no
rugæ; its parietes were very thin; and it fcarcely retained any veftige of the ring of the pylorus.
This latter circumftance, if it did not proceed from
original conformation, or from fome former difeafe,
might be owing. as the others were, to the frequent
diftenfions which muft have taken place from
drunkennefs, and alfo to the late diftenfion from
the included air. The gall-bladder was contracted, and contained a very little bile. The kidneys
were flabby; and the aorta was fmaller than ufual.
The ovaria were even fmaller and more fhrivelled
than they are at that age. A hydatid, of the bulk
of an ordinary fized grape, adhered to the left ovarium. The uterus was inclined to the right fide;
its fundus was rather fmall. The cervix, efpecially at the lower part, was thicker than ufual; and
the os uteri was fmall and of a circular form, as in
virgins. From this a fluid fimilar in colour and confiftence to milk, which was not fetid, and not in very fmall quantity, diftilled; from this it appeared

I   that

that the woman had laboured under fluor aibus.
The fource of this fluid was found to be higher
than the lower part of the cervix. At that part
the veficles of the cervix contained mucus, probably more fluid than ufual, but not like milk.
None however were obferved higher up. About
the middle of the cervix, a round empty cell, capable of containing a fmall French bean, was
found in two places buried within the fubftance
of the parietes, which exhibited no marks of ero-
fion either there or in any other place.

THORAX. The vertebræ of the thorax were bent
to the right fide much more than ufual, which
proved that the woman had been hump-backed;
a circumftance that had been indicated by the
lumbar vertebræ having appeared to have begun
to incline to the left fide, but fo flightly that the
innominata were not affected by it. No fluid was
found effufed in the thorax or pericardium. The
lungs anteriorly, and at the fides, were very ftrongly connected to the pleura; they were befides tu-
mid, and efpecially the left lobe, but only in confequence of air and a little fluid mixed with it.
They were no where indurated, nor of a redder
colour than natural. The ventricles of the heart
contained coagulated blood of a black colour, as
it was in the other parts of the body.

HEAD. All the the contents of the cranium
appeared more flabby than ufual, although they
were

were examined within eight days after death. The veffels of the pia mater were turgid with blood. A number of veficles filled with watery fluid, and not very fmall, were obferved in the choroid plexufes. All the contents of the cranium having been removed, that furface of both the petrous proceffes which is next the brain, and the parts contiguous to it, were found to be unequal, and not fmooth as ufual.

CAUSES of Continued Inflammatory Fever.

PREDISPONENT CAUSE. Although plethora, however induced, be evidently a neceffary predifponent caufe to this fever, it does not appear to be afcertained whether fome other circumftance be not alfo required.

EXCITING CAUSES. Sudden tranfitions from heat to cold, fwallowing cold drinks when the body is heated, exceffive exercife, intemperance, violent paffions of the mind, the fudden fuppreffion of habitual evacuations, the fudden repulfion of eruptions, and what the French call a coup de foleil, are probably the fole exciting caufes of this difeafe. That this fever never originates from perfonal infection is generally allowed; and the accounts of

its

its having occafionally appeared as an epidemic *
are too vague to be credited †, and even, although
proved, could afford no decifive evidence that it
was occafioned by contagion.

PROXIMATE CAUSE. Many modern phyficians
have confidered all fevers to be produced by the
fame proximate caufe ‡, and their opinion is now
very generally received. Waving for the prefent
the confideration of this fubject, it may be obferved, that the theories refpecting the proximate
caufe of fever which have of late years prevailed,
do not explain that of continued inflammatory
fever.

The firft of thefe theories is that of Boerhaave.
He imagines that, in confequence of lentor of the
blood, there is a ftagnation and refiftance in the
extreme veffels; while, at the fame time, the heart
is irritated into irregular action from an inordinate
motion of the nervous fluid into it §. Befides
which, he fuppofes that, during the courfe of the

VOL. I. D fever,

---

* Vide, An Enquiry into the Nature, Rife, and Progrefs of the
Fevers moft common in London, by Wm. Grant, M. D. page 193;
and Sydenham's Works, paffim.

† Vide, Collection d' Obfervations fur les Maladies et Conftitu-
tions Epidemiques, par M. Lepecque de la Cloture, pag. 836.

‡ Vide, The Works of Dr. Cullen; and Obfervations on Fevers,
by Dr. Clarke of Newcaftle.

§ Vide, Aphor. Boerhaavi, 581, 593; et Van Swieten Comment.
in locis.

fever, fuch a change in the fluids, as their becoming very thick, acrimonious, &c. may take place, as fhall vary the type of the difeafe *.

Several circumftances contradict this theory: *Firft*, An inordinate motion of the nervous fluid is fo vague an expreffion, that it is not eafy to underftand what is meant by it. *Secondly*, No fatisfactory proofs have been produced, to render it certain that the ftate of the fluids is altered at the beginning of inflammatory fever; while, at the fame time, many facts make it probable that the change in the blood is the confequence, and not the caufe, of the increafed action of the vafcular fyftem. And *Thirdly*, The determination of blood to the head, fo general in this fever, is not explained by this theory.

The fecond opinion was originally fuggefted by Hoffman †, and has fince been improved by Dr. Cullen. He alleges that, by the application of the remote caufes of fever, the energy of the brain is diminifhed, and debility of the whole of the functions, and particularly of the action of the extreme veffels, is produced : that the nature of the animal oeconomy is fuch, that this debility proves an indirect ftimulus to the fanguiferous fyftem; whence, by the intervention of the cold ftage and

fpafm

* Vide Aph. 592, 593.
† Hoffmanni opera omnia, tom. ii. pag. 10.

fpafm connected with it, the action of the heart and larger arteries is increafed ; and continues fo till it has had the effect of reftoring the energy of the brain, of extending the energy to the extreme veffels, of reftoring therefore their action, and thereby efpecially overcoming the fpafm affecting them ; upon the removing of which the excretion of fweat, and other marks of the relaxation of the excretories, take place *.

This theory does not explain the proximate caufe of inflammatory fever, for the following reafons : *Firft*, Becaufe, according to his own definition of the difeafe, the functions of the brain are little difturbed. *Secondly*, Becaufe debility is generally more confiderable at the end than at the beginning of the paroxyfm. And *Thirdly*, Becaufe in inflammatory fever the hot fit is violent, although the cold fit be flight ; whereas, were Dr. Cullen's theory true, the hot fit fhould always be in proportion to the preceding cold one.

Thefe theories being found thus inadequate to the explanation of the proximate caufe of this difeafe, it might be expected that fome other fhould be here fubftituted : but it is much eafier to overturn than to eftablifh theories. The following obfervations therefore are offered merely as fuggeftions, defigned only to afford a few imperfect hints.

D 2

The

* Vide Firft Lines, par. 46.

The propenfity of medical authors to attribute complex and contradictory phenomena to a fingle caufe, has been highly prejudicial to the progrefs of medicine. It is probable that the different fpecies of fevers proceed from different proximate caufes; and hence, in the invefligation of the caufe of each, the phenomena peculiar to each fpecies, and not thofe common to all, ought to be confidered. This is the more particularly neceffary, as all difeafes confift of primary and fecondary fymptoms; for no part of the body can be deranged without other parts being affected. As a preliminary ftep, therefore, to the invefligation of the proximate caufe of inflammatory fever, it is neceffary to afcertain the primary fymptoms. Thefe appear to be, violent action of the fanguiferous fyftem, and increafed heat of the whole body. All the other fymptoms feem to be fecondary. An exception indeed may be urged in favour of the cold fit; but as that fymptom is not only common to all fevers, but alfo fometimes perhaps is wanting in inflammatory fever, it cannot be regarded as a primary one. Befides, it may appear to fome, that increafed heat of the body fhould not be ftated as a primary fymptom, from its being probably the confequence of the increafed action of the vafcular fyftem. There is, however, fome ambiguity in this circumftance;

for increafed hert of the body, or at leaft of particular parts of it, is fometimes felt independent of increafed action of the fanguiferous fyftem. As the primary fymptoms of the difeafe cannot be very clearly afcertained, an attempt to explain thofe mentioned might not perhaps be deemed very fatisfactory.

It is well known that, in a ftate of health, expofure to any of the exciting caufes of inflammatory fever produces a train of fymptoms refembling thofe of that fever, which in a few hours fpontaneoufly ceafe by the eruption of fweat. To what circumftance then ought the permanent increafed action of the vafcular fyftem in inflammatory fever to be attributed? Can the plethoric ftate of the fyftem, previous to the application of the exciting caufe, account for the difference in the degree of violence and duration of thefe fymptoms in the two cafes?

With refpect to the fecondary fymptoms, that which it is moft difficult to explain is the cold fit. As for the reafons already ftated, it ought to be confidered as an acceffary, and not a neceffary fymptom: it cannot be regarded as depending folely on the action of the exciting caufes. That it depends upon the diminifhed action of the veffels on the furface of the body is very evident; but whether this proceeds from the action of caufes applied immediately to the fanguiferous fyftem
itfelf,

itſelf, or mediately through the nervous ſyſtem, is not eaſily determined.

That the deviation from nature in the appearance of the blood is the conſequence of the increaſed action of the veſſels has been generally admitted. Yet it is probable that it is not owing excluſively to that circumſtance.

The determination of blood to the head ſo obvious, not only from the ſymptoms, but alſo from the appearances on diſſection, deſerves particular notice, as it may perhaps afford a clue to the labyrinth in which the proximate cauſe of inflammatory fever has been hitherto concealed.

———

REMARKS ON THE CASES OF INFLAMMATORY FEVER.

THOSE caſes prove decidedly, in contradiction to the aſſertions of many medical practitioners, the exiſtence of inflammatory fever, as deſcribed in the preceding pages from the writings of authors.

The third, fourth, and fifth caſes, ſhow that this fever does not occur excluſively in perſons under forty years of age; and that it is not, as has been alledged by ſome, confined to cold climates, as all of them occured in Italy.

The eighth caſe affords a ſtriking illuſtration of

the

the obfervations of Lommius, refpecting the danger of yellownefs of the fkin, when it occurs
before the feventh day; and alfo refpecting the
tendency of this fever to terminate in inflammation of the lungs.

———————

§ 2. NERVOUS FEVER, or TYPHUS.

As this fever appears under a variety of forms,
it has been divided by authors into various fpecies. Unlefs this plan were adopted, an accurate
defcription of its fymptoms and types could not
be given.

I. Slow Nervous Fever, or Typhus Mitior *.
This does not, like the inflammatory fever, invade
fuddenly. The patient at firft feels liftlefs and oppreffed; his appetite for food is diminifhed; he is
reftlefs; has great dejection of fpirits; feels an
unufual wearinefs after the leaft motion; and
has alternate chills and fhiverings, with fudden
flufhes of heat. A lucid interval generally takes
place

* The fymptoms of this fever are detailed more accurately by
Huxham, p. 74, than by any other author. The defcription here
given is borrowed chiefly from him.

place in the morning, but all the fymptoms are aggravated at night; and this circumftance continues, in a greater or lefs degree, throughout the progrefs of the difeafe. After two or three days, vertigo, or pain of the head, efpecially about the hinder part, attended fometimes with a fenfe of coldnefs in the courfe of the coronary future; naufea, with the vomiting of infipid phlegm, and great proftration of ftrength, efpecially felt when the erect pofture is attempted, fupervene. At the fame time, the heat of the body is not confiderably increafed; the pulfe is quick, weak, and irregular; the tongue is moift, white, and covered with a vifcid mucus; the breathing is difficult, accompanied with oppreffion about the præcordia; the urine is pale, watery, and fometimes like whey; the belly is generally coftive. Infomnolency takes place; and although fometimes the patient appears to be afleep he is quite unconfcious of it. All thefe fymptoms are attended with great inability to exert the powers of the mind or of the body.

In this fituation the patient continues for fome days; the fymptoms then become aggravated; the face is fometimes hot and flufhed when the feet are cold, and at the fame time cold fweats on the forehead and on the backs of the hands break out. Tinnitus aurium takes place, and may be regarded as the forerunner of delirium. Deliquia

often

often occur, efpecially if the erect pofture be attempted. If at this period of the difeafe the urine
continue clear and watery, delirium with fubfultus tendinum enfue. The delirium is feldom violent, appearing at firft when the patient awakes
like the continuance of a dream; from which ftate
the patient recollects himfelf for a little; but the
confufion foon returns, and from this he cannot
for fome time be roufed. At laft he continues
in a conftant dozing muttering ftate. Profufe
fweats or colliquative diarrhœa fucceed; the patient is quite ftupified, and infenfible to the impreffions of both light and noife, although at
the beginning he had been remarkably fufceptible of fuch impreffions. Great exhauftion now
takes place; the tongue, which is dry, efpecially
in the middle, trembles when the patient puts it
out; the extremities are cold; the nails of the
fingers are livid; the pulfe flutters, is fo indiftinct
as fcarcely to be felt, and is eafily compreffible.
Trembling and twitching of the hands fometimes
occur, and are the preludes to ftrong convulfions
which fnap the thread of life. In other cafes, the
fenfes of feeing and hearing are completely loft;
the delirium is converted into coma; the ftools
and urine pafs involuntarily; fubfultus tendinum
fupervene; a particular noife in the throat, em-
phatically called rattling in the throat, is heard;
and the patient finks. This fatal termination moft

Vol. I.   E   generally

generally happens about the fourteenth day, feldom before the eighth or ninth, nor after the thirtieth.

The termination of this fever is very uncertain; and it is impoffible, at the beginning, to judge whether recovery or death fhall happen: For fometimes, after the fymptoms have been quite flight, a fudden aggravation takes place, and the patient is cut off; while, in other cafes, the moft alarming fymptoms, fuch as convulfions, coma, &c. precede a favourable change. In general, however, if the patient furvive the fourteenth day without any bad fymptoms, recovery may be expected.

Moift tongue, foft fkin with gentle fweats, the pulfe continuing firm or becoming ftrong, the urine depofiting a copious fediment, and more efpecially flownefs of the pulfe after gentle fweats or moderate diarrhœa, the appearance of the ftools becoming natural, and the appetite for food returning, are favourable figns. A pretty free falivation alfo without aphthæ, attended with kindly moifture of the fkin, and impofthumes about the ear or about the parotid gland, or a large, puftular and angry eruption about the lips and nofe, are regarded as fymptoms portending a favourable event.

On the contrary, where the delirium fupervenes early, and continues above four days; where there are profufe difcharges, either in the form of diar-

rhœa

rhœa or colliquative fweats, along with weak pulfe; where there is fubfultus tendinum, tremors of the hands or of the tongue, conftant infomnolency, blindnefs, impeded deglutition (which takes place, when it occurs, about the eleventh day) more efpe- cially if attended with fingultus; where the ex- tremities are cold, and livid petechiæ appear, much danger is to be apprehended. But, as already ftated, it is very difficult, or perhaps rather impof- fible, to form a decifive opinion refpecting the e- vent.

In moft cafes, gangrene, in a greater or lefs degree, occurs in thofe parts on which the body had principally refted during the progrefs of the difeafe. In general, the fores in confequence heal kindly, after the feverifh fymptoms have difap- peared; but fometimes it happens that the ulce- rations are fo very extenfive that the patient finks under the difcharge.

II. Malignant Putrid Fever, or Typhus Gravior *. In this fpecies, the fymptoms at the beginning are more violent than in the former. The rigors, if any take place, are more confide- rable; there is intenfe diftreffing heat, at firft re- mittent, afterwards permanent; the pulfe is hard, fmall, quick, and unequal; there is great proftra-

E 2   tion

---

* For an accurate defcription of the fymptoms of this fpecies, fee Huxham, p. 92; alfo Home, Princip. Medicin. p. 88.

tion of ftrength, and much anxiety and defponden-
cy. Great naufea and vomiting of black bile, violent
pain of the head and of the temples, throbbing
of the temporal and carotid arteries, tinnitus aurium, and laborious refpiration, interrupted by
fighing, attend; and at the fame time the breath
is fetid. The eyes are inflamed, and a pain is
felt about the orbits; the countenance feems
bloated, and has a cadaverous appearance. Pains
in the ftomach, limbs, and back, fupervene; and
alfo tremors and delirium. The tongue at firft is
white, afterwards becomes black and dry, fo as to
render the fpeech inarticulate; the lips and teeth
are covered with a black vifcid fordes; and there
is great thirft, with a bitter tafte in the mouth.
Sometimes, however, although the tongue and
fauces be remarkably parched, no thirft is felt:
in fuch cafes, phrenzy or coma always enfue.
The urine is at firft pale; but, during the progrefs of the difeafe, becomes very high coloured,
and fometimes black, with a very fetid foot-like
fediment. The ftools are intolerably ftinking;
of a green, livid, or black colour; and frequently attended with gripes, and with the difcharge
of blood. Small livid or red-coloured fpots, like
flea-bites, called petechiæ, or broad fpots of the
fame colour, called vibices, appear over the body, fometimes about the fourth or fifth day, and
fometimes not till the eleventh day, or even later.

An

An efflorefcence alfo like the meafles, but of a lefs bright colour, in which the fkin, efpecially on the breaft, appears as it were marbled and variegated, is in fome cafes obferved. Thefe eruptions are commonly attended or preceded by profufe fetid fweats. Sometimes, though very rarely, on the eruption of the fweat the petechiæ difappear, and fmall white miliary puftules break out: in other cafes an itching, fmarting, red rafh, or large fretting watery bladders on the back, breaft, fhoulders, &c. are obferved. White or dark coloured aphthæ appear in the mouth; and are foon fucceeded by great difficulty of fwallowing, pain in and ulceration of the fauces, the œfophagus, &c. together with inceffant fingultus. Hæmorrhages, fometimes from the nofe, but more commonly from the inteftines, as the ftools are bloody, and at the fame time fanious, black, and horribly fetid, then take place. Towards the end of the difeafe, where it terminates fatally, the petechiæ become of a green dark colour, or livid black vibices appear; and thefe, together with coldnefs of the extremities, are generally the forerunners of death. Blood drawn during the courfe of this fever has a livid appearance, has its component parts fcarcely cohering, and foon runs into putrefaction.

As in the former fpecies of typhus, fo alfo in this, no certain opinion refpecting the event of the difeafe

can

can be formed. The moſt favourable ſymptoms
are: a yellow or brown colour of the ſtools; the
fetid diarrhœa and ſweating not taking place till
a late period of the diſeaſe; and the petechiæ
changing from a dark to a bright red colour.
The bad ſymptoms are: no thirſt; numerous
black petechiæ, or the ſudden receſſion of the petechiæ, with a very feeble pulſe; livid aphthæ; inflammation of the fauces; laborious reſpiration after the appearance of a miliary eruption; ſwelling of the abdomen after profuſe ſtools; diarrhœa,
with the diſcharge of very fetid, bloody, ichorous
matter; coldneſs of the extremities; and convulſions.

The appearances on diſſection in thoſe who die
of this fever exhibit inflammation and often gangrene of the brain and other viſcera, more eſpecially the ſtomach and inteſtines *.

III. YELLOW FEVER †. This ſpecies of fever
occurs in America, and the Weſt Indies. It begins

* Vide Home Princip. Medicin. pag. 89; alſo the Tranſlation of
Hoffman, reviſed and corrected by Dr. Duncan, vol. i. page 169.

† For an account of this fever, ſee, A Treatiſe concerning the
Malignant Fever in Barbadoes, by Henry Warren, M. D. p. 9;
Rouppe de Morbis Navigantium, pag. 304; Blane on the Diſeaſes
of Seamen; Hillary's Obſervations on the Diſeaſes of Barbadoes;
Mackittrick Diſſertatio de Febre Indiæ Occide talis Maligna fla-
va; and an Inaugural Diſſertation on the ſubject by Sam. Cur-
tin in Webſter's Medicinae Praxeos Syſtema, vol. i.

gins with a fudden faintnefs, fometimes giddinefs,
fucceeded by flight fenfe of cold; fevere pain in
the head above the eye-brows; with a flight naufea, and impaired appetite. Intenfe heat foon
after fucceeds; the face is flufhed; the eyes are
impatient of light, are inflamed, and have a
painful fenfe of heat. The pulfe, in the mean
time, is full, quick, but foft; the carotid arteries
throb; while the pain of the head is fo violent,
that the patient feels as if the temples fhould be
rent afunder; pains are alfo felt in the joints, and
in the loins. The naufea becomes aggravated,
and vomiting is excited; and this ftate of ftomach
continues throughout the whole difeafe, every
thing that is fwallowed being commonly rejected.
The matter vomited has different appearances:
fometimes it confifts merely of what had been
fwallowed; at other times it is pure bile; and at
other times, it feems to be acrid bile. Straitnefs
and oppreffion about the præcordia accompany
the vomiting. The hypochondria, at the fame
time, generally become more or lefs fwelled; and
great forenefs to the touch is felt at the pit of the
ftomach. When the patient attempts the erect
pofture, he is giddy, and feels as if he fhould fall.
Obftinate infomnolency takes place; or, if he
have a fhort fleep, the patient awakens in a fright.
Great thirft and a bitter tafte in the mouth are
felt. The tongue, at the beginning of the difeafe,

appears

appears in fome cafes fhining; in others white;
and on very rare occafions, it is of a yellow colour, and covered with fordes. The belly is for
the moft part bound; but fometimes diarrhœa
takes place at the beginning. The urine is high
coloured, and is fecreted in lefs quantity than
ufual. In this manner the difeafe begins. Within thirty-fix or forty-eight hours, however, that
is generally in the morning of the fecond day after the attack, a deceitful remiffion happens, that
impofes upon the patient and attendants. But
within a few hours the vomiting becomes more
violent than formerly; and the matter vomited is
commonly black. Exceffive thirft is felt. The
tongue is rough and brown in the middle; but
below, and at the fides, it is, together with the
gums and lips, of a very florid red colour:
towards the end of the difeafe, it frequently appears black. Thefe fymptoms are commonly aggravated towards the evening, and are fucceeded
by a very reftlefs night. On the third day,
though fometimes fooner and fometimes later, a
clammy fweat, that is by no means refrefhing,
breaks out; the pulfe becomes flow, languid, and
foft; the fkin is fenfibly cold; and an appearance
of relief again takes place. But, inftead of relief,
the ftrength of the patient is more and more impaired; a conftant anxiety continues; the refpi-
rations become lefs frequent, and the breathing is

3   loud.

loud. Deliquium is induced by the flighteft caufes, principally when the patient attempts to raife
himfelf in bed. The face and neck become yellow inftead of red; but the rednefs again returns
for a fhort time. The watchfulnefs is more conftant; the patient is fometimes flightly delirious,
and again becomes fufficiently recollected. The
rednefs of the eyes is firft changed into a brown,
and then into a yellow colour; with which, in
a fhort time after, the whole external furface
of the body is tinged. The vomiting is not fo
troublefome upon the fourth day. Putrid bile,
however, mixed with black blood, is vomited, and
is paffed alfo by ftool, but is not attended with
gripes. Blood in a diffolved ftate is effufed, fometimes from the nofe, and fometimes from the gums.
The pulfe becomes fmall, fluttering, and intermitting; and almoft conftant fingultus attends. The
hands and feet are cold, and at the fame time
fwelled, and of a purple colour. The lips are
parched, and covered with a livid fordes; the u-
rine is very yellow, and depofites a fediment which
is almoft quite black. Coma now fupervenes:
and in fome cafes the breathing is like that in
apoplexy, now ftertorous, and now eafy; in other
cafes coma is very flight, and is interrupted by delirium. Matters remain fometimes in this fituation for twenty-four or thirty-fix hours, the patient lying like one almoft dead, and fcarcely any

pulfation being felt at the wrift. In the mean time broad livid fpots appear about the præcordia and loins. Violent convulfions at laft come on, which terminate in death.

The event of this fever is always very precarious. It often proves fatal within twenty-four hours after its attack. Sometimes the fatal termination is protracted to the eleventh or twelfth day; but it moft commonly happens between the fourth and feventh. Where the patient recovers, there is no regularly marked crifis. The bad fymptoms are: extreme weaknefs from the beginning; early black vomiting, as, before the fourth day; early yellownefs of the fkin; hæmorrhages from the nofe, lungs, or urinary paffages, after the third day; and livid blotches about the præcordia and loins. The favourable fymptoms are: firmnefs of the pulfe; no extreme proftration of ftrength; no black vomiting, nor yellownefs of the fkin, till after the fifth or fixth day; and, on fome rare occafions, a large eruption of boils over the whole body.

In addition to the above hiftory, the following circumftances, detailed by Dr Rufh, as having characterifed the yellow fever which appeared at Philadelphia in 1793, deferve attention.

The pulfe was, both at the beginning and during the courfe of the difeafe, exceedingly irregular. At the beginning, although often full, tenfe, and quick,

it

it was frequently fo low as fcarcely to be felt at the
wrifls; or intermitting, or preternaturally flow. Dilatation of the pupil generally occurred in cafes
where it was flow; but the flownefs of the pulfe
ufually preceded the dilatation of the pupil. After the fever had continued for fome weeks, and
the weather had become more cool, (that is, after
the tenth of September) the pulfe was as full,
tenfe, quick, and frequent, at the beginning, as
in cafes of pleurify; at the fame time, however, it
communicated, when felt, a peculiar fenfation to
the fingers, no two pulfations being exactly fimilar. It was equally full, hard, and frequent, in
the remiffions of the fever as in the exacerbations. Although before death it commonly became
weak, frequent, and imperceptible; yet in feveral
cafes it was full, active, and even tenfe, during the
laft hours of life.

Many complained of a dull pain in the region
of the liver; but few of that forenefs to the touch
at the pit of the ftomach noticed in all former hiftories of the yellow fever. A burning pain, however, in the region of the ftomach, accompanied
the vomiting which occurred about the fourth or
fifth day. The appearance of what was vomited
was different at different ftages of the difeafe. On
the firft and fecond days, it confifted commonly of
pure bile. About the fourth or fifth day, it refembled coffee impregnated with its grounds. To-

wards

wards the clofe of the difeafe, it was of a pale black colour, and feemed to be acrid bile with a flaky fubftance floating in it ; and at the very clofe it was dark-coloured grumous blood. Along with thefe difcharges of the ftomach, there was a large worm fometimes, and often tough mucus.

The ftools varied in appearance according to the treatment of the difeafe. They generally indicated a fuperabundance of bile; but in fome cafes they were as white as in jaundice. The difcharge of urine was fometimes accompanied by a burning pain, refembling that which takes place in gonorrhœa. A total deficiency in the fecretion of urine, without any pain, occurred in many cafes for a day or two.

Tremors of the limbs, and twitchings of the tendons, previous to the fatal termination of the difeafe, were uncommon. In fome cafes, a morbid degree of ftrength to a wonderful extent took place at that period. In the greateft number of inftances, the patients died in a placid manner.

The difeafe appeared under a variety of forms. It was often fo mild that the patients were not confined to their beds. In fuch cafes, the only fymptoms were, fallownefs of the countenance, naufea, univerfal languor, and irregularity of the pulfe. It appeared invariably to affect chiefly the weak parts of the fyftem which it attacked; as the head, the lungs, the ftomach, the

2   bowels,

bowels, and the limbs, suffered more or less according as they were more or less debilitated by previous inflammatory or nervous diseases. Soon after the fever became generally prevalent, every other disease seemed to yield to its superior influence; and hence a number of anomalous symptoms occurred, in many cases, which imposed on several practitioners the belief that in such cases the fever did not exist.

The appetite for food returned much sooner during the course of convalescence after this disease than it does in ordinary fevers *.

WHERE death happens, the process of putrefaction advances so rapidly, that it is often necessary to inter the body within a few hours after the fatal event. The appearances on dissection † exhibit the stomach, intestines, and mesentary, covered with gangrenous spots. The orifice of the stomach seems greatly affected, the spots upon it being ulcerated. The liver and lungs are said to appear also of a putrid colour and texture.

Dr Rush remarks, that the appearances on dissection were different in different cases; and seemed to be owing to determinations of the fluids

to

---

* Vide, An Account of the Bilious Remitting Yellow Fever, as it appeared in the city of Philadelphia in the year 1793; by B. Rush, M. D. p. 40 et seq.

† Vide Lind's Essay on the Diseases of Europeans in Hot Climates, p. 114.

to different parts *. Unequivocal fymptoms of morbid congeftions of blood in feveral of the vif-cera, but more efpecially in the brain, appeared, he obferves, in many cafes †.

IV. MIXED CONTINUED FEVER, or SYNOCHUS. Cafes frequently happen where the fymptoms of fever do not occur in the order of fucceffion mentioned under the three preceding heads. Thofe fevers, according to Dr Cullen, generally confift of fuch a complication of the fymptoms of inflammatory and nervous fever, that it is at the beginning difficult to afcertain the fpecies to which they belong; but that at firft the fymptoms are thofe of inflammatory fever; and afterwards, fometimes gradually and fometimes fuddenly, they degenerate into thofe of typhus.

Fevers of this kind appear frequently in Scotland. Although on their firft attack they be attended with violent pain in the head, ftrong hard quick pulfe, flufhed face, intenfe heat, fometimes not preceded either by languor or by a cold fit; yet the naufea, giddinefs, and diminifhed energy of the brain, which at the fame time take place, together with certain circumftances refpecting their exciting caufes, to be afterwards mentioned, clearly fhow that the difeafe is to be referred to the

* Page 114.   † Page 48.

the fpecies of typhus, as the event of the cafe fooner or later proves.

FROM the above defcription of the various fpecies of continued nervous fever it appears, that the effential characters of typhus are, impaired energy of the brain, confiderable proftration of ftrength, irregular action of the vafcular fyftem, and a deranged ftate of the chylopoetic vifcera. This definition, however, is neither fo accurate nor fo fatisfactory as could be wifhed; yet it is not eafy, nor perhaps in the prefent imperfect ftate of phyfiology poffible, to form a better one from the hiftory of the fymptoms alone.

Typhus therefore differs from fynocha in one of the great effential characters; namely, impaired energy of the brain. It differs too in feveral other particulars refpecting the fymptoms; as, in being preceded by languor and liftlefsnefs, in being attended with a weak pulfe, &c. and in being protracted to a longer period.

The feat of the difeafe in typhus has not yet been clearly afcertained. Every part of the fyftem is affected; the powers both of the body and mind are deranged; and it is difficult to determine whether one part be affected before the others, and whether the derangement of one part depend upon or be unconnected with that of others.

CASES

CASES of Continued Nervous Fever.

I. Slow Nervous Fever.

Case I. (xlix. 2.)

A man, of about thirty years of age, was affected with a flow fever; which was accompanied with no fymptom deferving notice, except that his appetite for food had entirely failed. His pulfe and ftrength became every day weaker; and at length he died fuddenly.

Appearances on Diffection.

Abdomen. The bile in the gall bladder was tinged of a brown colour. A calculus of the fize of a dens molaris, of a pale colour, and very friable on its furface, was found in it. This gall ftone contained feveral other fmall ones of a black colour.

Thorax. The lungs were marked with black fpots. The pericardium contained little or no ferum. The blood, which in other parts of this body had an unnatural dirty appearance, was in the ventricles of the heart in a coagulated ftate.

Case II. (vii. 6.)

A porter, labouring under an acute fever, was affected with a very violent pain in his head, which

was

was fucceeded by delirium. Soon after which he died.

Appearances on Diffection.

Head. A fmall quantity of ferous fluid was found between the dura and pia mater; part of it, coagulated like tranfparent jelly, was feen among the fanguiferous veffels. In the finus of the falx a long flender concretion was obferved. The whole brain had the natural appearance.

C a s e III. (xlix. 12.)

A man, aged forty years, was received into the hofpital of Bologna, in confequence of a wound in the tibia occafioned by a blow with a bludgeon. When his wound had affumed a favourable appearance, and he himfelf was in good health, he was fuddenly affected with an acute fever; which gradually increafing in violence, at laft terminated in death.

Appearances on Diffection.

No uncommon appearance whatever was difcovered, except that the blood retained nearly the natural vital fluidity.

C a s e IV. (i. 14.)

A woman, who was previoufly affected with lues venerea, having been feized with fever, at-

Vol. I.   G   tended

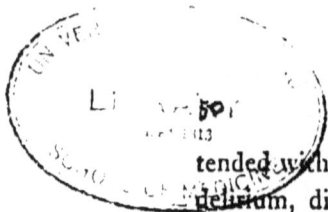

tended with excruciating pains in the head, and delirium, died in the hofpital of Padua.

Appearances on Diffection.

HEAD. The fkull internally appeared in fome places of a blackifh red colour. That portion of the dura mater next the fuperior middle part of the right lateral finus was much thickened; and it coalefced perfectly with the pia mater, and alfo with the fubftance of the brain. The meninges and brain in that part were almoft femiputrid, and tainted with a dirty colour of a yellowifh cineritious hue: this colour was moft confpicuous in the cortical fubftance of the brain. When the cerebellum was drawn out from the hollow formed by the dura mater, its external furface was fo clofely connected to the dura and pia mater that part of it adhered to them; but the adhefion did not exceed in extent two fingers breadth, and therefore it was not fo confiderable as in the cerebrum. The finufes of the dura mater, and the veffels of the pia mater, which were larger than ufual, were diftended with black blood. The fanguiferous veffels, in many places throughout the medullary fubftance of the brain, appeared to be very numerous, and were more diftinct than they commonly are. The lateral ventricles were filled with a fluid of a brown colour, and their furfaces were tinged with the fame colour. The pineal gland was firmer, larger, and whiter than ufual, and feemed

ed

ed to contain a kind of cells. This woman had a peculiarity in the form of her fkull which de- ferves notice ; the right fide projected more out- wards at the pofterior part than the left, and con- fequently the cavity of that fide, and the hemi- fphere of the brain contained in it, were larger than thofe of the left fide.

C A S E V. (XIV. 5.)

A YOUNG man affected with fever was received into the hofpital of Bologna. He had, it appear- ed, an old fiftula above the right maftoid procefs. Injections thrown into it returned partly by the neighbouring ear, with which neverthelefs he heard. The fever having increafed in violence, within a few days after his admiffion he became delirious and comatofe ; and in that ftate died.

Appearances on Diffection.

HEAD. All the veffels of the brain were tur- gid. There was a confiderable quantity of green ferous fluid in the lateral ventricles; and from that contained in the right ventricle fome pus of a greenifh colour fubfided. A much greater quan- tity of the fame kind of pus was obferved between the dura mater and the internal and inferior fur- face of the petrous procefs of the right temporal bone; and thus paffing between both it extended into the vertebral tube. The cavity of the tym-

panum was alſo filled with the ſame kind of matter. That ſurface of the petrous proceſs, on which
are placed the Fallopian duct and the ſemicircular
canals, was ſo much eroded, that a pretty wide
cleft appeared on the poſterior ſide of the foramen,
through which both portions of the auditory nerves
paſs. That cleft was covered with the dura mater, which there ſeemed to be alſo eroded, but
not to ſuch an extent as the cleft itſelf. Although
the brain had a proper degree of firmneſs, and was
examined the day after death, the ſmell was ſo
offenſively putrid that it was not poſſible to trace
the paſſage of the matter from the cleft to the
right ventricle, nor yet to aſcertain whether any
other parts within the ear beſides the tympanum
were injured.

Case VI. (xxi. 9.)

A husbandman, aged thirty five years, having
fallen from a great height, and bruiſed his right
ſide, was received into the hoſpital of Padua. This
man, after he had recovered ſo completely that he
was about to be diſmiſſed from the hoſpital, having by ſtealth crammed himſelf with bread and
ſome other indigeſtible aliment, was ſuddenly affected with violent fever, to which difficulty of
breathing ſoon after ſucceeded. There was, however, no pain in the cheſt. Although thoſe remedies

dies which the diforder feemed to require were employed, efpecially blood-letting once or twice, all
was to no purpofe; for he died in eight days after
the attack. During the latter days of his life, he
lay on his back, was affected with a ftertor in his
breathing, and had a very frequent, for fome time
full, and latterly, cord-like pulfe.

Appearances on Diffection.

ABDOMEN. The large inteftines were particularly confpicuous, being much diftended with air.
The lower parts of the ileum were to a confiderable extent of a blackifh red colour; the blood veffels being as diftinct as if they had been filled with
coloured wax. There was alfo a ftrong fmell from
the parts, fuch as is commonly felt in inflamed inteftines. The liver was externally of a whitifh colour, with livid edges; internally, it was variegated
like marble. The gall-bladder contained a fmall
quantity of bile only; and the bile was not perfectly natural. The fpleen was large, whitifh, and
very flabby.

THORAX. The lungs adhered every where to
the pleura, not only by means of membranes, but
apparently alfo immediately by themfelves. When
however this was examined more accurately, it
was found that a thin yellow concretion, refembling a membrane was interpofed between the
membrane of the lungs (which was found) and
the pleura. When the lungs were drawn forward,

both

both that concretion and the pleura itfelf followed.
The inferior portion of the left lobe was very
large, heavy, much indurated, and refembled
the fubftance of the liver. The other parts of the
lungs were not unfound. The pericardium contained a quantity of yellowifh turbid fluid. Polypous concretions were found in the large veffels
and auricles of the heart. The largeft of thefe
concretions was obferved in the right auricle; its
branches extended from thence into the jugular
veins.

C A S E VII. (LV. II.)

A PORTER, of a tall ftature, robuft, although
pale in the face, and of a lean habit of body, aged
about forty years, addicted to drinking, had been
affected with fcabies, for which he had taken fulphur mixed with wine. After having continued
for fix months in apparent good health, he was
feized with fever, and brought into the hofpital of
Bologna.

On the firft day, the ftate of the pulfe and the
other fymptoms were not very bad; but his fituation was rendered doubtful by a certain anxiety
and reftlefsnefs. On that day a medicine was giv-
en with a view to cleanfe out gently the ftomach
and bowels. On the day following, he was affected with vomiting and difficulty of refpiration, and
was

was convulfed in his whole body; he was befides violently delirious, and in his delirium cried out that his belly was on fire. Under thefe circum-ftances he died, on the third day after his admiffion into the hofpital.

Appearances on Diffection.

ABDOMEN. The fat (which was not in large quantity) contained in the omentum, in the pin-guedinous appendiculæ of the colon, and efpecial-ly in thofe that were near the fpleen, and alfo in the adipofe membrane of the left kidney, was of a brown and livid colour. The colon was diftended to three times its ufual fize by the contained air. The ftomach was contracted, was of a brown co-lour externally; and internally, at that part which is next the fpleen, was inflamed, without any ero-fion, to the extent of an hand-breadth. The low-er part of the liver was of a livid blackifh colour, but the appearance was quite fuperficial.

THORAX. The lungs, the heart, and the dia-phragm, were found in a natural ftate. The ventricles contained fome polypous concretions.

CASE VIII. (XLIX. 14.)

A NOBLEMAN, aged about forty years, who when in health was fubject to fuch conftipation of the bowels as required the conftant ufe of purgative medicines, removed his habitation from his native country

country, the air of which was pure, to a fituation
where the atmofphere was denfe. Soon after this
he was affected with fever, marked by no other
fymptom than a continual anxiety of mind and
conftant ftate of watching. On the fourteenth
day pain in the head fupervened; and his pulfe
became fo irregular that it could only be felt now
and then. In this fituation he died.

Appearances on Diffection.

ABDOMEN. The fundus of the ftomach was
tinged with a black colour. The inteftines, and
efpecially the large ones, were contracted. The
other abdominal vifcera were found. All the
blood was ftrongly coagulated.

CASE IX. (IV. 6.)

A MAN of a flender make, about forty years of
age, became affected with acute fever. During
the night of about the ninth day, he loft the fa-
culty of fpeech. When fpoken to, he fhowed no
fign of underftanding what was faid. A flight
power only of moving and feeling remained in
all his limbs. His face was not red. About the
thirtieth day he died.

Appearances on Diffection.

HEAD. Some ferous fluid was found effufed
between the meninges of the brain; the ventricles

were

were full of a fimilar fluid. Every thing elfe was
natural.

C A S E X. (v. 2.)

A MAN, aged thirty-three years, of a fanguineous temperament, of a lean habit of body,
affected with hernia, and much addicted to the
ufe of wine and tobacco, began to complain of
pain in the left fide of his head, particularly about
the occiput, which was followed by a pain and
weaknefs of the mufcles of the neck on the fame
fide. Thefe fymptoms were accompanied with
fever, which was at firft violent, but afterwards
feemed to remit. His pulfe became not only
flow, but alfo weak, affording little refiftance to
the fingers; and at the fame time, his ftrength
began to fail, fo that every motion of his body
now became difficult. After interrupted delirium, he loft the power of fpeech, and of motion;
and at laft, finking gradually, he died on the fourteenth day of the difeafe.

Appearances on Diffection.

ABDOMEN. The inteftines within the fcrotum
forming the hernia, were fo ftrongly connected by
furrounding membranes to the tefticle, that they
could not be replaced.

HEAD. When the brain was taken out of the
fkull, a little purulent matter was obferved at its

bafis; and when that was wiped away, on hand-
ling the brain more matter appeared. This proceeded from the ventricles through the infundi-
bulum; for both left and right ventricle, but
more efpecially the latter, contained a confiderable quantity of that matter. There was a foramen in the corpus ftriatum of that ventricle, which
communicated with a finuous ulcer occupying the
third part of the fubftance conftituting the bafis of
the brain on the right fide. No morbid appearance occurred in the left fide of the brain.

CASE XI. (VII. 2.)

A YOUNG man, about twenty years of age, who
had a flow fever, attended with thirft, was admitted into the hofpital of Bologna. As the appearance of his urine, and the ftate of his pulfe, were
in fome meafure fimilar to thofe of perfons in
health, the difeafe feemed flight. About the eighth
day, however, violent delirium fupervened, which
continued for feven days. The delirium having
abated, he remained in a ftate of imbecility, except that at fome times his recollection returned.
At laft he died.

Appearances on Diffection.

EXTERNALLY. The body approached fomewhat
to a livid appearance, efpecially under the nails

of

of the fingers; and the mufcular flefh was rather of a brownifh than of the natural red colour.

THORAX. All the vifcera were found; there, however, was no appearance of fluid in the pericardium.

HEAD. A gelatinous concretion was obferved furrounding the fanguiferous veffels which creep through the pia mater. That membrane, at the bafis of the fkull, being torn, a quantity of fluid, which in colour and confiftence refembled cow-milk whey, flowed out. The whole brain was found. The blood in this body was of a black colour, and was thick, but ftill fluid.

CASE XII. (VII. 4.)

A MAN, aged thirty-five years, became affected with pain in the thorax, accompanied with fever. The pain having ceafed, delirium fupervened; while the fever conftantly increafed in violence, and continued till death. This event took place on the eleventh day.

Appearances on Diffection.

THORAX. The pofterior parts of the lungs were fomewhat hard, and were of a red colour. A polypous concretion appeared in each ventricle of the heart; the larger of which was in the left ventricle. This is an unufual circumftance, the larger being commonly found in the right.

H 2　　　　　　　HEAD.

HEAD. When the brain was taken out of the cranium, a little ferous fluid was difcharged from the meninges. A polypous concretion appeared in each of the large finufes of the dura mater. The whole brain was foft. The plexus choroides of the left ventricle was very turgid, and, as it were, varicofe.

CASE XIII. (XVI. 6.)

A MAN, about forty years of age, after having laboured under a flow fever for many weeks, had a flight fwelling of his feet, a tickling cough, and parched fauces. His refpiration was fhort and difficult, and required that his neck fhould be erect. His pulfe could be fcarcely felt. He died fuddenly.

Appearances on Diffection.

ABDOMEN. Although a watery fluid was difcharged from the left fide of the umbilical region, when the teguments of the abdomen were in the courfe of diffection feparated from the mufcles; yet every thing in that cavity was in a natural ftate, except the fpleen, which was three times larger than ufual.

THORAX. A limpid ferous fluid was found ftagnating in each fide of the cheft. This being put into a glafs veffel, depofited at the bottom of the veffel fome matter in feparate portions, fo that it

had

had no continued membranous fubftance floating in it, as ferum in the thorax generally has.
The lungs were found, and entirely free, except
that the lower part of the left lobe was conneéled
to the pleura by a fhort and flender membranous
band. The pericardium was dilated, and contained more than half a pound of limpid ferous
fluid. The heart was large, and in its right ventricle efpecially, a flaccid polypous concretion was
feen, a circumftance which is rare in cafes where
water is found ftagnating in the cavities. The
thoracic duét, and alfo the lymphatics in the abdomen, were fo empty, that not the leaft veftige
of them could be traced.

C A S E XIV. (xxi. 15.)

An old man, aged ninety, who had been for
fome time in the hofpital at Padua on account of
a contufion on his thigh, became, without any apparent caufe, affeéled with flight fever. No remarkable fymptoms occurred. His pulfe was quick
and weak, but never intermitting. Within a few
days he gradually funk.

Appearances on Diffection.

ABDOMEN. The aorta and iliacs had here and
there bony fcales. The orifice of the pylorus was
furnifhed with what is called its valve at one part
only. In all the remaining part, which was by
                                    much

much the largeſt portion, there was no veſtige of
that kind, and no appearance of its ever having ex-
iſted. The orifice too was much larger than uſual.
In the ſubſtance of the left kidney there were cavities in two places; in one the cavity was ſmall, in
the other large. Theſe cavities contained a watery fluid included within the proper coat of the kidney, which was extended over the upper part.
The urinary bladder, found in other reſpects, grew
out at the left ſide above the inſertion of the ureters into a cell compoſed of its own coats, of a he-
miſpherical figure and of a moderate ſize. It had,
by a ſmall opening, a communication with this
cell. In the ſcrotum a hernia was obſerved, which
ſeemed to have formerly contained a portion of
the inteſtinal canal, or at leaſt a larger portion of
the omentum than it then held. It was a pretty
large ſac, deſcending almoſt as far as the teſticle
from the ring of the oblique muſcle on the right
ſide between the membrane attached to the cre-
maſter muſcle and the tunica vaginalis, which was
apparently found. This ſac lay on the inſide of
the ſpermatic veſſels. It was formed by the peritoneum having fallen down through an orifice
capable of admitting the finger, and then having
become dilated and thickened. A ſmall and ſlender fold of the omentum had paſſed through that
orifice into the ſac, and adhered firmly to its pa-
rieteş.

THORAX.

THORAX. A fmall quantity of red coloured watery fluid was found in each fide of the thorax; the left lobe of the lungs was connected at fome places to the pleura, which was perfectly found ; but the right lobe had almoft no adhefions. This latter lobe, however, at the lower part, was fwelled and indurated from inflammation. Almoft the whole anterior furface of the heart was covered with fat, although the man was in other refpects lean. The valves, at the orifices of the right ventricle, were not only neither rigid nor thickened; but even appeared, as did alfo thofe that belong to the pulmonary artery, to be compofed of rather a thinner membrane than ufual. In the left ventricle, however, the mitral valves were thicker than ordinary; and befides, the femilunar were entirely bony and inflexible. Internally they protuberated fo, that they were at a diftance from the parietes of the artery. They alfo grew out into a fmall thick body in the middle of their border. The aorta from the heart, as far as that part where it adheres to the vertebræ, as well as the carotids and fubclavians, exhibited no bony fcales; although fuch fcales were evident to the touch throughout the remaining courfe of the aorta within the thorax.

CASE

CASE XV. (VI. 4.)

A WOMAN of about twenty five years of age became affected with malignant fever, attended from
the beginning with deafnefs. On the feventeenth
day coma fupervened, fo that, when fpoken to, fhe
neither anfwered, nor opened her eyes. In this fituation fhe died.

Appearances on Diffection.

HEAD. The brain appeared to be quite found
in every refpect, except that, when it was taken
out of the fkull to be examined, a fmall quantity
of ferous fluid was difcharged through the infun-
dibulum. In the cavity of the tympanum of the
ear, and in the neighbouring finufes, fome fanious
matter was found.

II. TYPHUS GRAVIOR.

CASE I. (X. 5.)

AN unmarried woman, aged twenty fix years,
affected with acute fever, was about the feventh day
feized with convulfions, fo that when fpoken to fhe
laughed in that convulfive manner ftiled rifus far-
donicus. The convulfions were attended with delirium, and were fo violent, that it was neceffary to
tie her in bed that fhe might not fall out. All

2   thefe

thefe fymptoms ceafed an hour before her death, which happened on the ninth day. Her refpira-ration had become more and more difficult.

Appearances on Diffection.

EXTERNALLY. A little ferum had flowed out from the right ear.

THORAX. In the right fide of the cheft, the lungs adhered to the fternum, and laterally to the ribs; and externally, on that part of them next the clavicle, a certain fubftance of an inter-mediate nature between fat and a gelatinous con-cretion, fuch as fometimes floats on the fluid of dropfical patients, was found. This circumftance rendered it probable that this woman had labour-ed under fome diforder of the thorax previous to that difeafe which had proved fatal. Eight or nine ounces of ferous fluid were found effufed with-in the cavity of the thorax on the fame fide. Po-lypous concretions like mucus were obferved in each ventricle of the heart; that in the right ven-tricle was the largeft.

HEAD. The whole brain appeared found; and nothing remarkable occurred within the cranium, except that when the dura mater was in fome pla-ces torn, in feparating it from the fkull, fome drops of ferum flowed out; and blood much diluted with ferum was difcharged from fome very minute veffels which were accidentally lacerated.

C a s e XXI. (xxxi. 2.)

A young man who, during the whole courfe of his life, even in the beft health, had his bowels in a loofe ftate, having reached his twentieth year, became affected with gripes in his belly, attended with frequent bloody ftools, or in other words had dyfentery. After twelve or fifteen days, his difeafe was converted into a fimple diarrhœa, with ftools of a yellow colour, and unaccompanied with gripes. When, by means of proper medicines the diarrhœa had become fomewhat alleviated, he was feized with a tertian fever, from which within a month he was relieved. The diarrhœa having ftill continued, he was fuddenly attacked with acute fever, that had frequent exacerbations. His pulfe was foft, fmall, weak, and quick. To thefe fymptoms confufion of mind, and a peculiar fwelling of the anterior part of the left fide of the thorax, fupervened. Thus affected, he died, about the fourteenth day from the beginning of the acute difeafe.

Appearances on Diffection.

Abdomen. Although the belly did not feem in the fmalleft degree fwelled, it contained a great quantity of fanious ichor that had proceeded from the inteftinal canal, a certain portion of which was perforated in feveral places. That portion

comprehended

comprehended the extremity of the ileum, and alfo the contiguous part of the colon, to the length
of two hands breadth. The canal was there eroded, ulcerated, and on its internal furface, gangrenous, fo that the perforations were not to be wondered at. Near this difeafed part of the inteftines
fome of the glands 'of the mefentery had grown
out into a tumour, containing ichor not unlike
that found in the cavity of the belly. The fubftance of that tumour was foft and flabby, and appeared approaching towards a ftate of putrefaction.
The fpleen was three times larger than natural.

THORAX. When the fkin and mufcles of the
thorax, covering the fwelling formerly mentioned,
were cut into, a large quantity of ferous fluid was
difcharged, efpecially at the fide of the upper part
of the fternum ; for there ferum ran out in ftreams
from the borders of the pectoral and fubclavian
mufcles. The lungs were found. The pericardium contained ferum, like water in which frefh
meat had been wafhed. The heart was fo foft
and flabby, that, when felt with the fingers, it
feemed to be membranous rather than mufcular.
In the ventricles fluid blood, which was fo frothy
that it might have been compared to foap fuds,
was found. In all the veins there was fuch a
quantity of air, that, although they had very little
blood in them, they were exceffively diftended.
This was more efpecially the cafe with one large

I 2   branch

branch of the veins which belong to the fpleen; for, although it appeared as if it could not be dilated to a greater extent, fcarcely any veftige of blood could be traced in it.

HEAD. A fmall quantity of ferous fluid appeared within the cranium; but the brain itfelf exhibited no marks of difeafe.

CASE III. (XLIX. 24.)

A WOMAN, aged fifty years, affected with malignant fever, was admitted into the hofpital of Padua. Her pulfe during the firft fix days was fmall and indiftinct, but afterwards became fomewhat more perceptible. A fenfation of confiderable ftraitnefs of the breaft during refpiration, together with palpitation of the heart, having fupervened, fhe died within the fpace of two days.

Appearances on Diffection.

THORAX. The blood in the left ventricle of the heart was half coagulated. In the right ventricle, which it diftended, it formed a polypous concretion, apparently of a flefhy confiftence; but neverthelefs fo tenacious, that it refifted the knife as much as the moft vifcid cruft found on the blood of pleuritic patients ufually does.

III. MIXED

III. Mixed Fever.

Case I. (xlix. 6.)

A woman, aged thirty years, of a bilious temperament, who had for a confiderable time laboured under a double tertian fever, received a blow upon the abdomen with a ftick. Having been admitted into the hofpital of Padua, fhe complained only of pain in the abdomen; but on the third day after the blow fhe began to be delirious. Her pulfe was fmall, and quick; and fhe vomited more than once a fluid refembling water in which frefh meat had been wafhed. At length, the difeafe having increafed every day, fhe died.

Appearances on Diffection.

ABDOMEN. The mufcles were found to be contufed; in fuch a manner, neverthelefs, that no marks of contufion appeared, neither on the outfide nor on the infide of the belly. The liver was of a whitifh colour, and fomewhat harder than ufual. The gall-bladder was of a very large fize; and contained about three ounces of bile of a black colour. Some of the fame kind of bile was found in the ftomach. Every thing elfe within the abdomen was ftrictly natural.

THORAX. The lungs, on the furface which was

turned

turned towards the vertebræ, were confiderably inflamed; but in other refpects found. The right
ventricle of the heart contained a fmall polypous
concretion.

CASE II. (XVI. 17.)

A GIRL of about fifteen years of age was affected with acute fever, attended principally with a
violent pain in the head, for the other fymptoms
were mild. The fever feemed to remit about the
tenth day. Within a few days, however, great
thirft, laborious refpiration, and pain in the left
fide of the thorax, fupervened to the fever. Thefe
two latter fymptoms having continued to increafe
in violence, her fpeech and underftanding being unimpaired, fhe died within a few days, contrary to the expectation of thofe who attended
her.

Appearances on Diffection.

THORAX. The lungs were found; but the left
cavity of the cheft was filled with fluid not very
unlike the urine of horfes: in this fome concretions, refembling the white of an egg, floated.
There was alfo a very fmall quantity of ferous
fluid in the right fide of the cheft. The pericardium was completely filled with a fluid of a thicker confiftence than that in the thorax; and as the
external furface of the heart was flightly eroded,

it

it was probable that it had been fo eroded by the
fame fluid. Polypous concretions, fomewhat like
condenfed mucus, appeared in the ventricles of
the heart. That in the left ventricle was the larger.

CASE III. (xxiv. 6.)

AN old man who had been for three months in
the hofpital of Bologna on account of a fractured
leg, was detained there from a flight, though obftinate fever. At laft it was unexpectedly difcovered that he had no pulfe, although it had not
hitherto intermitted, and although the man af-
ferted that he did not feel himfelf worfe than
ufual at that time. With a view to confirm what
he faid, he immediately raifed himfelf up, and fat
erect in bed. In a fhort time, however, he died.

Appearances on Diffection.

ABDOMEN. One of the kidneys contained under its proper coat a cell, of the fize of a fmall
bean, filled with ferum. Granules, as if of tobacco, appeared at the fides of the feminal caruncle.
Within the pendulous part of the urethra only
one of the lacunæ, and that a fmall one, was obferved.

THORAX. There was no dilatation in any part
of the thoracic vifcera or large veffels; although
the aorta exhibited, near its valves and in other
places,

places, marks indicating future offification; and although it had a fmall bony lamella within its internal coat at that part where, after having fent off the left fubclavian, it began to defcend. Polypous concretions were here and there feen in the ventricles of the heart, efpecially the right one, and in the large blood veffels. One of thefe was of a white colour; and of fo firm a texture, that it gave great refiftance when an attempt was made to pull it in pieces.

HEAD. Although no watery fluid flowed out from the vertebral tube when the upper cervical vertebræ were feparated from the lower, a confiderable quantity was difcharged during the opening of the head, and the lateral ventricles contained not a little. The fubftance of the brain was not flabby, but was even perhaps the hardeft that was ever felt. The blood veffels, fo far from being pale, were moft of them (efpecially the finufes) filled with black blood. Although the weather was cold, and it was fcarcely three days fince the death of the fubject, and although there were no marks of putrefaction in any part of the body, air bubbles were feen in thofe two arteries which pafs between the hemifpheres of the brain, near the fuperior furface of the corpus callofum. A confiderable portion of the parietes of the right carotid artery, near the cavernous finus, had become thickened, and of an intermediate nature

2   between

between ligament and cartilage; yet on its internal furface it was even membranous. A fimilar morbid appearance feemed begun in other large veſſels of the brain. On the outſide of the cranium that difeafed ſtate of the veſſels was much more confiderable; for on one ſide of the neck a pretty large bony fcale was found between the coats of the carotid, juſt at its diviſion, and the whole trunk of that artery was very much dilated. The other carotid was of the natural dimenſions.

C A S E IV. (xxxvi. 23.)

A wool-comber, of about forty years of age, came into the hoſpital at Padua, on account (as he himſelf faid) of obſtructions in the hypochondria. The bad colour of his face; the bad health he had fuffered for a whole year; the irregular fever with which he was often troubled, and which was not abfent at that time; and more efpecially the examination of both hypochondria, particularly the right one, confirmed what he had faid. When he feemed to have received fome benefit from the ufe of the medicines prefcribed, he became unexpectedly affected with acute fever, accompanied with fymptoms of internal inflammation of the thorax; and within ten or twelve days died.

Appearances on Diffection.

ABDOMEN. Although the weather was cold, and it was not yet two days fince the death of the fub-ject, the abdominal mufcles were flabby, and at the lower part were of a greenifh colour. The liver was uncommonly large; and, although externally its colour did not appear bad, internally it was of a pale brown colour: befides, when accurately exa-mined, the whole of it both internally and exter-nally was marked with certain brown fpots, and was harder than ufual; a circumftance which was afcertained both by the fingers and by the knife, as it was cut into in various directions and in diffe-rent places. It was obferved that no yellow point appeared from the fections of any of the veins in cutting into the liver, which is the ufual mark of the hepatic ducts being alfo cut into. The gall-bladder was fmall in proportion to the fize of the liver, and contained a little bile of a colour ap-proaching to that of afhes, fo that it was uncertain whether the hepatic ducts had become collapfed from the fmall quantity of bile in the gall-bladder, or if that pale colour of the bile itfelf had pre-vented the appearance of the above-mentioned mark of thofe veffels having been cut through. The fpleen, in all its dimenfions, was double the natural fize; but in other refpects was found. The fplenic artery, from its origin to its termination, was no where tortuous, or fo to fpeak, varicofe as

ufual,

ufual, except in one place, about the middle of its courfe, where it was fomewhat inflected. A hard body, of the bulk and almoft of the figure of a middle fized cherry, but with an unequal granulated furface, of an intermediate nature between bone and ftone, had grown on the mefentery. Clofe by one fide of this body, an arterial and venous branch paffed, but did not enter its fubftance, and from thence proceeded to the inteftines, which lay at the diftance of about two fingers breadth from that body. The ileum, at the fide next the mefentery, was in one part fo inflected as to form an angle, which continued even after the mefentery was cut off; and, at the fame place on the oppofite fide, grew out into a fhort appendix of an hemifpherical form. The left external iliac vein, near the opening of the internal one, was hard but not offeous; for the coats had at that part for a fhort tract become thickened only. Thefe coats being opened, fmall chords, and fome fmaller fubftances like valves, appeared in the cavity of the vein on one fide where it was not perforated by any orifices. The kidneys were about nine inches in length; but were proportionally narrow, except at the fuperior extremity, where they were a little broader. The length of the finufes that receive and tranfmit the veffels was alfo uncommon, which was the more readily obferved, as all that part of the fubftance of the kidneys that fhould have

made

made the anterior paries of the finufes was wanting. The larger of thofe branches which convey
the urine into the pelvis of the kidney were therefore wholly expofed, as were alfo the fanguiferous
veffels in all that part generally concealed within
the finus. Two arteries, the one fuperior and the
other inferior, and as many veins, belonged to each
kidney. The veins went out of the finufes in fuch
a manner that the inferior one afcended in an oblique direction into the fuperior, which was placed tranfverfely. The arteries, however, formed
no junction with each other; both the inferior
and fuperior extended tranfverfely, without any
obliquity; and, confequently, the inferior did not
enter the finus, but paffed below it almoft at the
lower fide of the kidney. Both the inferior arteries had their origin much lower than the renal
arteries ufually have, as they proceeded from the
aorta, at the diftance of fcarcely an inch above its
divifion into the iliacs; befides, they did not come
out from the fides of that veffel, but from the very middle of its anterior furface, and were fo near
to each other that their orifices were feparated by
a very thin feptum only. From that part of the
aorta, they proceeded on each fide, both of them
being fimilar, and without having any ramifications before their infertion; and were diftributed
over the kidneys in the manner already defcribed.
The fuperior arteries, on the other hand, which
were

were fomewhat thicker than the inferior ones, differed neither in their origin nor ramifications from
the real natural renal arteries.

THORAX. The right lobe of the lungs adhered
to the pleura, and was indurated. Some fpoons-
ful of bloody watery fluid were found in the pericardium. Two veins, which ran longitudinally
upon the pofterior furface of the heart were turgid
with blood, and as it were varicofe.

CASE V. (XVI. 38.)

A COUNTRY woman, not above twenty-five years
of age, of a pale countenance, who had been married for about four months, and who was three
months pregnant, having become affected with a
flight irregular fever, was admitted into the hofpital
of Padua, where fhe lay for a month or more. Her
pulfe was neither fmall nor intermitting, although
fhe almoft lived entirely upon fruit; which was
not to be wondered at, confidering fhe was pregnant. She had no thirft. She had no fwelling
in the feet; nor was fhe affected with faintings.
She never complained of any ftraitnefs or uneafinefs about the præcordia, nor of any fenfation of
weight, nor of any other uneafy fymptoms in the
thorax, except a flight dry cough, to which fhe had
been occafionally for a long time habitually fubject. Although her refpiration became difficult

if

if she swallowed any thing warm, and for that rea-
son she begged to have every thing cold, yet she
had in other respects no trouble in breathing : for
during the night she never had any sense of suffo-
cation, nor was ever obliged to sit up in bed on account of her breathing. She lay on her right side.
In this position, although no symptom had supervened to the slight fever, except a pain in the loins,
of which she had complained in the very last days
of her life only, she died.

Appearances on Dissection.

ABDOMEN. The belly was opened within half
an hour after the death of the mother, with a
view, according to the custom of the country, to
baptise the child if it were alive. This object was
accomplished ; for, on cautiously cutting through
the uterus and membranes, the infant rushed out,
as it were, and moved its hands. It did not cease
to live till an hour after the death of the mother.
The spleen was a little larger than natural. The
liver was much more so ; as it extended both lower down than ordinary, and also across quite to
the spleen. Externally it was of a pale colour ;
and internally it appeared variegated with its own
proper colour and that of tobacco. The stomach,
almost in the middle of its length, was contracted,
and at each extremity was swelled, but in a less
degree at the right side. which part descended in
an oblique direction. The other extremity was
placed

placed tranfverfely, with its fundus turned fomewhat anteriorly; and was half full of air and fluid.
A great number of lumbrici appeared in the fmall
inteftines; and wherever they were, but efpecially in one place, the inteftines were of a red colour,
and protuberated as if forced outwards by violent
means.

THORAX. Before the thorax was opened the
neck was obferved to be fwelled, from the tur-
gefcence of the thyroid gland; and it was found
that milk could be eafily preffed out from the breafts.
In the right cavity of the thorax there was a great
quantity of yellowifh watery fluid; in which fome
thick mucus, and as it were membranous fubftances, appeared. Some of the fame kind of fluid was
found alfo in the left fide; and there was fo much
of it in the pericardium, that it was almoft completely filled with it. In it too the fame membranous fubftances were feen. Polypous concretions,
formed of a kind of mucus, were found in each
ventricle of the heart: thofe in the left ventricle
were a little thicker than thofe in the right. A
very fmall quantity of fluid appeared at the extreme parts of the feet; although no marks of it
had been perceived, neither during life nor before
diffection. The mufcles all over the body were
in the moft natural ftate.

CASE.

C a s e V. (xxi. 6.)

A mason, aged about thirty years, became affected with fever, in confequence of fatigue in
working. To this periodical daily rigors fupervened; and foon after he began to be delirious
during the time of the rigors. At length the di-
lirium, which ufed to go off immediately after the
rigors, continued conftantly; and from that time
he grew worfe and worfe. The delirium was of
the melancholy and plaintive kind. His pulfe
was not irregular. Blood was fometimes difcharged from the noftrils. Although venefection was
had recourfe to even three times, and other modes
of cure which feemed neceffary were employed;
yet, as he continued always to grow worfe, he gradually at laft died.

Appearances on Diffection.

Abdomen. The fpleen was large.

Thorax. The lungs, except at their anterior
part, which was of a whitifh colour and in a found
ftate, were almoft every where harder than natural. The right lobe, efpecially at the upper part,
was exceedingly indurated, very heavy, diftended, of a red colour, and compofed of a ftrong thick
fubftance. Although the body was not opened
till nine hours after death, all the vifcera were ftill
warm and fmoking; and the blood which flowed

2   from

from their veins when divided, was fluid and warm. Polypous concretions neverthelefs were drawn out, not only from the crural veins, but alfo from the heart, from whence they extended to the pulmonary veffels of one fide at leaft.

CAUSES of Continued Nervous Fever.

Predisponent Cause. Many facts concur to prove, that a certain ftate of the fyftem is required for the action of the exciting caufe or caufes of this fever. The particular nature, however, of that ftate, has not hitherto been afcertained. That it depends upon the condition of the nervous fyftem, there is every reafon to believe; but our ignorance of the laws by which that fyftem is regulated, prevents us from approaching nearer to a folution of the queftion. Perfons of all ages, and of both fexes, are fubject to the difeafe. While fome appear wonderfully fufceptible of it; others, though conftantly expofed to the exciting caufes, are not affected by them: thus, for example, medical practitioners and nurfe-tenders are feldom infected with fever; and criminals from a jail have been known to communicate contagion to numerous perfons in a crowded court, although they themfelves were in good health.

Vol. I.   L   Exciting

EXCITING CAUSES. The variety of circumſtances which have been commonly regarded as exciting cauſes of nervous fever, has tended much
to perplex thoſe who have attempted to inveſti-
gate the nature of this ſpecies of fever. Expoſure to cold, as it is termed; diſorders in the primæ viæ; intemperance; violent paſſions of the
mind; fatigue of mind or body; corrupted animal exhalations; exhalations from the putrefaction
of vegetables; a particular ſtate of the atmoſphere,
from the viciſſitude of ſeaſons, or the ſucceſſion
of weather; confined human effluvia; and contagion, have been enumerated as the exciting
cauſes.

1. *Expoſure to Cold.* This has been reckoned by the common bulk of mankind as a frequent
exciting cauſe of fever; a circumſtance to which
perhaps may be attributed its being ranked as
ſuch by medical practitioners. It may be thought
preſumptuous to call in queſtion an opinion, founded upon obſervation, which is ſo generally received. When, however, it is recollected, that expoſure to cold produces ſynocha, or inflammatory
fever, which is a diſeaſe very oppoſite in its nature
to typhus, as has been already proved; and that,
beſides, it frequently induces catarrh, rheuma-
tiſm, bowel complaints, &c. it muſt be obvious,
that ſome additional circumſtance at leaſt muſt
concur

concur before it can excite nervous fever. And if it be allowed that the former effects certainly do occur much more frequently than the latter has been fuppofed to do, it may probably appear, that expofure to cold ought to be expunged from the lift of exciting caufes of the difeafe under confideration.

2. *Diforders in the Primæ Viæ.* Thefe are more probably the confequences than the caufes of fever: for a difordered ftate of the ftomach and bowels frequently takes place, without being attended with nervous fever; whereas that fpecies of fever, unaccompanied with derangement of thofe vifcera, never occurs. This caufe, however, deferves great attention; for the ftate of the ftomach and its connections has certainly a remarkable influence over the whole fyftem. That the ordinary functions of the ftomach are not only deranged, but even totally fufpended, in nervous fever, is more than probable. Dr. Fordyce, for example, mentions that he has often feen cafes where food, taken before the attack of fever, had remained in the ftomach for three or four days *. But whether an impaired, deranged, or fufpended ftate of its ordinary functions, be the only morbid change that happens in the ftomach during fe-

L 2   ver,

* See A Differtation on Simple Fever, by George Fordyce, M.D. page 66.

ver, is a queſtion which cannot be eaſily determined. From the circumſtance already mentioned, that diſorders in the primæ viæ, impeding the
funƈtions of digeſtion, frequently occur independent of fever, it appears that ſome change not
yet explained certainly happens; and it is upon
this ſuppoſition that the author of theſe remarks
rejeƈts diſorders of the primæ viæ as an exciting
cauſe of nervous fever.

3. *Intemperance.* The many inſtances which
are obſerved where intemperance of diet is not
ſucceeded by nervous fever, render it more than
probable that, in thoſe few caſes where it is ſo,
ſome other cauſe concurs. The immediate morbid effeƈt of intemperance, when it appears under the form of fever, ſeems to be ſynocha, and
not typhus.

4. *Violent Paſſions of the Mind, and Fatigue.*
Although fear, grief, and fatigue of mind and
body, have been long conſidered as exciting
cauſes of nervous fever *; there is reaſon to believe that they rather contribute to produce the
ſtate of the ſyſtem which prediſpoſes to that diſeaſe. Where violent paſſion, ſuch as exceſſive
anger, is immediately followed by fever, it is
probable

* A very late author is of the ſame opinion; vide Dr. Currie's
Account of the Climate and Diſeaſes of the United States of America, page 122.

probable that the difeafe produced is fynocha.
The depreffing paffions, fuch as fear, &c. which
fometimes occur about the firft attack of fever,
are perhaps the effects of the peculiar ftate of the
fyftem that had taken place at that time.

5. *Corrupted Animal Exhalations.* The exhalations from marfhes generally proceed from
the putrefaction both of animal and vegetable
matter, yet they induce intermittent fevers only.
With refpect to the effects of exhalations from putrid animal matter alone, nothing certain can be
urged ; but, upon the whole, it appears more probable that they contribute to render the effects of
fome other exciting caufe more violent, than that
they are themfelves exciting caufes.

6. *Exhalations from Putrefaction of Vegetables.*
Dr Rufh * has adduced the moft unequivocal proof
that, on fome occafions, vegetables in a ftate of
putrefaction act as exciting caufes of malignant fevers.

By the daily obfervation of mankind, however,
it is clearly eftablifhed, that this is neither a general nor a common effect of vegetable putrefaction ;
and, confequently, where it does take place, there
muft be fome concurrence or combination of other
circumftances not yet perfectly afcertained. The
heat

---

* Vide Dr. Rufh's Account of the Yellow Fever, already referred
to, p. 153. et feq.

heat of the climate feems obvioufly indeed to contribute much to this effect; but it is probably not
the fole circumftance.

7. *A particular State of the Atmofphere, from
the Viciffitudes of Seafons, or Succeffion of Weather.*
The numerous hiftories of epidemics on record *
have induced many practitioners to believe, that
a particular ftate of the atmofphere fometimes
proves the exciting caufe of continued nervous
fever. This opinion, however, feems founded
on partial obfervation only; for it is more confiftent with facts, to fuppofe that a particular
ftate of the atmofphere only tends to render the
fyftem more fufceptible of the impreffion of exciting caufes. Thus, for example, we are told,
that under the fame atmofphere two difeafes of a
different nature prevailed univerfally at the fame
time †.

. 8. *Confined Human Effluvia.* That thefe excite fever there can be no doubt; the moft in-
contefiible

---

* Vide, Willis opera. Sydenham opera, paffim. Morton opera. Halleri Difputationes, vol. v. Mertens Obfervationes Me-
dicæ de Febribus Putridis, paffim. Cleghorn's Obfervations on the
Epidemical Difeafes of Minorca. Mofely on Tropical Difeafes,
page 122. Collection d' Obfervations fur les Maladies et Conftitu-
tions Epidemiques, par M. Lepecque de la Cloture, pag. 836.

† Cleghorn on the Epidemical Difeafes of Minorca, page 136.
Lind on Difeafes incident to Europeans in Hot Climates, p. 126.

contestible evidence of the fact has been adduced *.

9. *Contagion.* All the species of nervous fever are contagious; a circumstance which forms the most essential mark of distinction between these fevers and all others. In many cases the contagion can be distinctly traced. In others, although it cannot be ascertained, there is no certainty that it did not exist; for such is the nature of contagion, that it may be communicated not only from a diseased person to a healthy one, but also may be carried about the cloaths, &c. even of a healthy person. In a crowded city, therefore, it is impossible for any individual to be certain that he has not been exposed to contagion. Many practitioners allege that contagion is not a general exciting cause of nervous fever; because it does not, like other contagions, as that of small pox, &c. lose the power of again infecting the same individual. But this argument is equally strong against the idea of fever being produced, in any instance, by infection; a fact, however, which is too well established to be denied. There is no inconsistency

* " At the Lent assizes in Taunton, 1730, some prisoners who " were brought thither from Ivelchester goal infected the Court; " and Lord Chief Baron Pengelly, Sir James Sheppard, Serjeant, " John Pigot, Esq. Sheriff, and some hundreds besides, died of " the goal distemper." See The State of the Prisons in England and Wales, by John Howard, F. R. S. p. 18.

cy in imagining, that the different fpecies of contagion are regulated by different laws, as it is found
that they produce different effects. With refpect
to the nature of the contagion of fever, nothing
but conjecture can be offered; which is not furprifing, fince that of the contagion of fmall pox,
although it can be made the object of our fenfes,
which that of fever cannot, properly fpeaking, be,
is ftill unknown. From the fimilarity of their effects, it is probable that it coincides, in fome effential qualities, with human effluvia. Perhaps, alfo,
the peculiar fmell which proceeds from human effluvia, and from the bodies of thofe affected with
nervous fever, being nearly of the fame kind, affords an additional argument in favour of that conjecture.

Upon the whole, it is prefumed, that human effluvia, putrid exhalations under certain circumftances, and contagion, are the fole exciting caufes
of continued nervous fever; and that expofure to
cold, diforders in the primæ viæ, intemperance,
violent paffions of the mind, fatigue of mind or body,
corrupted animal exhalations, and a particular ftate
of the atmofphere from the viciffitude of feafons
or the fucceffion of weather, act only by inducing
that ftate of the fyftem which renders it fufceptible of the impreffion of the exciting caufes.

An inveftigation of the exciting caufes of difeafe
has been regarded as ufeful in two points of view:

3   *Firft*,

Firft, In order that means may be fuggefted for the prevention of difeafes; and *Secondly,* For the purpofe of afcertaining the proximate caufe. The firft of thefe views alone, indeed, has been generally avowed; but the pains which moft fyftematic writers have taken to explain the mode of action, or the effects of exciting caufes, clearly fhow that they have been actuated by the fecond view alfo. If future obfervation fhall eftablifh the validity of an opinion lately propofed by Dr Fordyce, that a fever produced by any caufe, like a body put in motion by an impulfe, will continue, although that caufe be no longer applied * ; it will then appear neceffary only to enquire into the exiftence of exciting caufes, and not into their mode of operation. As Dr Fordyce has not yet produced the evidence on which his opinion is founded, it cannot be regarded as unfair, if affent be refufed to the doctrine as applying generally to all fevers. It is a well known fact, for example, that intermittent fever often refifts obftinately every means of cure, and yields only in confequence of a removal from expofure to the exciting caufe.

PROXIMATE CAUSE. The modern opinions refpecting the proximate caufe of nervous continued fever may be arranged under five heads.

VOL. I.   M   1. The

---

* Vide Fordyce's Differtation on Simple Fever, page 1;1.

1. The introduction of a morbid matter into the fyftem.

2. Lentor of the blood.

3. Impaired energy of the brain, with fpafm of the extreme veffels.

4. Simple debility.

5. An over proportion or accumulation of carbon and hydrogen, and confequently a diminution of oxygen, and an exhaufted ftate of irritability.

1. *The introduction of a morbid matter into the fyftem.* This is a very old opinion, and, among the vulgar, a very common one. The principal modern authors who have adopted it are, Sydenham, and Dr Balfour. Dr Sydenham imagines that a materies febricalis is introduced into the blood; and that a commotion is therein excited, by which the matter is either feparated and expelled, or the blood itfelf is changed into a new ftate *. Dr Balfour's opinion is, that the contagious matter being conveyed into the ftomach and bowels, infects the mucus lining the inteftines, which being abforbed and mixed with the blood produces the febrile ftate †.

Several

* Vide The Works of Dr Sydenham, tranflated by Dr Wallis, vol. i. chap. 4.

† See A Treatife on Putrid Inteftinal Remitting Fevers, by Francis Balfour, M. D. p. 16. et feq.

Several ſtrong objeƈtions, founded upon the progreſs and event of fevers, may be urged againſt
this theory; but theſe it is unneceſſary to adduce:
for, although the alleged alteration in the ſtate of
the blood were proved, the introduƈtion of morbid matter into the ſyſtem could be conſidered as
an exciting cauſe of fever only, and not the proximate óne. No change, however, of the blood,
has hitherto been demonſtrated as taking place invariably in every caſe of fever. On the contrary,
numerous faƈts concur to render it more than probable, that, ſo far as our ſenſes can determine, the
only changes which blood undergoes. in fevers of
this kind, are thoſe which proceed from the action of the blood-veſſels, and the deranged ſtate of
the chylopoetic viſcera; and hence ſuch changes
are the effeƈts, and not the cauſes of fever.

2. *Lentor of the Blood.* This opinion, propoſed
originally by Boerhaave, has been already explained, page 25. The objeƈtions already urged againſt
it, page 26, ſufficiently evince its inconſiſtency
with faƈts.

3. *Impaired Energy of the Brain, with Spaſm of
the extreme veſſels.* This theory has been alſo already explained, page 26. That it is inadequate
to the explanation of the phenomena of continued
fever, is preſumed from the two following circumſtances.

1ſt, Continued fevers are not always preceded

M 2 by

by a cold fit, nor by the other fymptoms regarded
as denoting fpafm of the extreme veffels.   And,

2*dly*, The energy of the brain is not always reftored on the ceffation of the fever ; for imbecilli-
ty of mind, which often continues for a confiderable time, is the frequent confequence of that difeafe.

Thefe objections to the doctrine of fpafm are fe-
lected as being incontrovertible.   Others, founded upon the inconfiftency of the feveral parts
or principles which conftitute the theory, might
have been urged.

4. *Simple Debility*.   This theory fcarcely re-
quires a fingle argument to prove its infufficien-
cy, to any one who confiders for a fingle mo-
ment the morbid ftates of the human body.   If
the debility which occafions fever be a particular
degree only in a fuppofed fcale between ordinary
health and the termination of life, then every
dropfical patient fhould be affected with fever:
and if it be a particular fpecies of debility ; that
is to fay, debility attended with peculiar circumftances, then the theory is merely a play upon
words.

5. *An over Proportion or Accumulation of Carbon
and Hydrogen, and confequently a Diminution of Oxygen, and an exhaufted State of Irritability*. This theory, founded upon the modern improvements in
chemiftry, has been lately propofed by Dr. Wood of
Newcaftle.

Newcaftle *. He proceeds upon the fuppofition
that a certain proportion of oxygen muft neceffarily
be received during every infpiration into the fyftem,
and that when this does not happen, carbon and
hydrogen gaining afcendancy, a putrefcent ftate
of the fyftem takes place.

An infuperable objection occurs againft this theory; viz. That the diminifhed proportion of oxygen
muft depend on fome peculiarity of action of the
fyftem in the perfon affected. For, as all mankind
have accefs to the fame grand magazine of oxygen, the atmofphere; (to ufe Dr. Wood's own
expreffion) no individual could be affected with
fever, while others efcaped, unlefs the reception
of oxygen, or the modification or diftribution of
the air we breathe, did depend upon corporeal
mechanifm †. Allowing, therefore, what is by no
means

* Thoughts on the Effects of the Application and Abftraction
of Stimuli on the Human Body, by James Wood, M. D. &c.
p. 60.

† Dr. Wood, in fupport of his theory, very properly adduces
practical fuccefs in addition to hypothetical reafoning. Proceeding upon the principle, that the acid of nitre, combined with pot-
afh, taken into the ftomach, communicates to the fyftem the oxygen which it contains; he regards the invariable fuccefs that attended the adminiftration of a folution of nitre in a great many
cafes of typhus, both under the care of his father and himfelf,
as a complete demonftration of the infallibility of his principles.
When, however, it is confidered, that the folution (confifting of
a drachm of nitre in eight ounces of water) was given in dofes
of

means proved, that an accumulation of carbon and hydrogen, with the confequences ftated, do actually happen in fever; they are to be regarded as effects, and not as caufes.

UPON the whole, it is obvious, that none of the modern theories above enumerated afford a rational explanation of the proximate caufe of continued fever; and it may be prefumed that, from the complication of morbid fymptoms which occurs in that difeafe, no fuch explanation can be given, until the laws, by which the nervous, the lymphatic, and the fanguiferous fyftem are regulated, and the mutual dependence that each has upon the other, be better underftood.

REMARKS ON THE CASES OF CONTINUED NERVOUS FEVER.

THE moft general appearance on diffection obferved in the preceding cafes is, effufion within the cranium. Yet in two cafes, viz. the fifteenth of typhus mitior, and the firft of typhus gravior,

every

of an ounce or fometimes two ounces every two or three hours; it may be eafily underftood, by one acquainted with the elements only of chemiftry, that before the half of the folution could be exhibited it muft have loft all power of giving out oxygen.

every thing within that cavity was found : confequently the moft general appearance feems to be
an effect which does not invariably take place.

With refpect to the different morbid appearances in the cavities of the thorax and abdomen, in
the cafes under confideration, they are to be regarded as accidental circumftances, only depending upon caufes, to be afterwards explained.

SECT. II. *INTERMITTENT FEVERS.*

INTERMITTENT FEVERS * are thofe where a perfect remiffion takes place, for a longer or fhorter
time, between the paroxyfms. Each paroxyfm
confifts of a regular cold, hot, and fweating fit ;
the phenomena of which are the following At
firft the patient complains of weaknefs, attended
with yawning and ftretching of his limbs ; followed

---

* For a defcription of intermittent fevers, fee, Boerhaave's Aph.
749, 750, 751 ; and Van Swieten's Commentaries on thefe. Confpect. Junckeri, pag. 651, et feq. Oofterdyk Prefcripta Medicinæ,
pag. 61. Cleghorn on the Difeafes of Minorca, page 147. Sydenham, tranflated by Wallis, vol. i. page 75. An accurate defcription of the general fymptoms of each paroxyfm is given by Dr.
Cullen, par. x. et feq.

ed by a certain fenfation of uneafinefs in the back
and in the points of the fingers. Thefe fymptoms
are fucceeded by fhivering, and a fenfe of great
cold ; although in fact the body is warmer than
natural. Naufea, vomiting; pain of the limbs,
of the back, and of the head; and difficult and
anxious refpiration, then take place. The pulfe
at the fame time is quick, feeble, fmall, and wire-
like ; and can be fcarcely reckoned on account
of the tremors of the body. The urine is limpid.
This ftage lafts for one, two, or more hours.

The cold fit having gradually ceafed, moft intolerable heat is felt ; the pulfe becomes full, ftrong,
and hard ; the refpiration lefs difficult, but ftill
anxious. The tongue is white, attended with
great thirft ; and the fenfation of exceffive heat is
felt about the præcordia. Headach, and fometimes
delirium, fupervene; and alfo pain, and in fome
cafes fwelling, about the region of the ftomach.
The naufea and vomiting often continue during
this ftage. The urine is of a red colour. Blood
drawn during this ftage is commonly thicker than
ufual; containing a fmall proportion of ferum,
and having a lefs firm cohefion of its particles
than natural. In fome cafes the upper part of
the craffamentum is red, and the under part black.
After the hot fit has continued for one, two, or
more hours, a copious fweat breaks out over the
whole body. All the fymptoms then become al-

3   leviated,

leviated, and after the fweat has lafted for three or four hours totally difappear. The urine depofites a fediment. Sleep takes place; and the infermiffion is completely eftablifhed, a fenfe of weaknefs only remaining. After a certain interval, in fome cafes longer, in others fhorter, the fame phenomena again recur in the fame fucceffion.

Authors have divided intermittents into different fpecies, according to the length of time interpofed between the beginning of one paroxyfm and that of another. Thus: where a regular fit of the difeafe takes place every twenty-four hours, it is named a Quotidian; if every forty-eight hours only, a Tertian; and if feventy-two hours intervene between the beginning of one fit and that of another, it is named a Quartan. Befides thefe, other divifions have been adopted, founded upon the irregularity of recurrence of the fits; as, the Semitertian, Double Tertian, Triple Tertian, &c. As intermittents occur moft frequently during the Spring and Autumn, they have alfo been divided into Vernal and Autumnal; and it has been found that tertians and quotidians moft frequently prevail in the former feafon, and quartans in the latter. Not contented with thefe divifions, authors have diftinguifhed intermittents from each other according to the anomalous fymptoms with which they are fometimes complicated. Accord-

ingly,

ingly, as the difeafe is occafionally attended with
violent fixed pain in the bowels, with faintings,
with fymptoms of apoplexy, or with pains in the
liver, fpleen, or kidneys; it has by different authors been ftyled, Affodes Syncopalis, Apoplectica,
Hepatica, Splenetica, Nephralgica, &c. Thefe
diftinctions, however, feem to ferve no ufeful purpofe; tending rather embarrafs than to inftruct.
It may be queftioned too, how far the divifion,
founded upon the intervals between the beginning of one paroxyfm and that of another, is of
practical utility; feeing, that during the courfe of
the difeafe the intervals frequently become protracted or fhortened : fo that what was originally
a quotidian fhall degenerate into a tertian or quartan, and what was at firft a tertian fhall be converted into a quotidian.   Sometimes too the paroxyfms fucceed each other fo rapidly, that no
regular intermiffion occurs; there being only a
temporary alleviation of the fymptoms, or remiffion as it has been called.   Such cafes have been
named Remittents.   For thefe reafons, it is impoffible to afcertain, at the firft attack of an intermittent, the exact form which it will affume :
for although, as already ftated, tertians and quotidians are moft frequent during Spring, and quartans and remittents prevail principally during Autumn; and although when the fit begins at noon
the difeafe moft generally affumes the tertian form,

and

and when in the evening the quartan type, there are fo many exceptions to thofe rules, that no dependence whatever can be placed upon them.

Intermittents are much more dangerous in warm climates than in temperate regions. In Great Britain the difeafe in many cafes terminates fpontaneoufly after a few fits, and leaves the patient nearly in a ftate of perfect health. In other cafes it refifts obftinately every remedy; and induces fo great a degree of weaknefs throughout the whole fyftem, that chronic difeafes of an alarming nature, depending upon vifceral obftructions, are induced. It often happens too, that, after the difeafe has continued obftinately for a confiderable length of time, it ceafes entirely upon the removal of the patient from the local fituation or place of refidence in which he had become affected with the difeafe; provided the change be made into what is called a dry atmofphere. In warm climates, on the contrary, not only do intermittents terminate in difeafes which become rapidly fatal, as apoplexy, dyfentery, &c. but often alfo do they prove immediately fatal. In fome cafes death takes place during the cold fit; but moft generally it happens during the hot one. Where the patient is not carried off by the firft attack, the fever becomes commonly irregular previous to its fatal termination. The favourable fymptoms commonly enumerated are: *Firft*, Regularity in the fta-

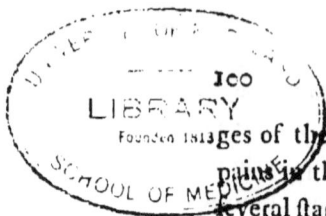

stages of the fit; *Secondly*, The abfence of violent pains in the bowels during the fits; *Thirdly*, The feveral ftages of the fit not occupying above twelve hours; *Fourthly*, The patient being, with refpect to the animal and natural functions, nearly in a ftate of health during the intermiffions; *Fifthly*, The urine depofiting a fediment; and *Sixthly*, The eruption of puftules about the lips on the apparent declenfion of the difeafe. The unfavourable fymptoms, on the contrary, are: *Firft*, Irregularity in the periods of attack, and in the occurrence of the ftages of the fit; *Secondly*, The fits becoming about the third or fourth period of the difeafe much protracted, the intermiffions confequently being greatly fhortened; *Thirdly*, Coma, delirium, great anxiety, painful fwelling and induration about the hypochondria and epigaftric region, pain about the upper orifice of the ftomach or loins, occurring during the fits; *Fourthly*, The patient, during the intermiffions, having a total averfion for food, and feeling fuch weaknefs, attended with vertigo, that he can fcarcely walk; *Fifthly*, The frequent appearance of numerous blotches on the fkin like the ftinging of nettles; *Sixthly*, The urine continuing thin, clear, high coloured, or being covered with an afh-coloured membrane like a cobweb; *Seventhly*, Hæmorrhages from the nofe, vomiting, colliquative fweats, or diarrhœa, taking place; and *Laftly*, The patient lying con-

ftantly

ftantly on his back, with a ghaftly countenance,
eyes half fhut, and mouth open; while at the fame
time the belly is fwelled to an enormous fize, and
obftinate coftivencfs attends, or involuntary difcharge of the feces happens *.   Intermittents frequently terminate alfo in continued fevers.

The appearances on diffection, in thofe who die
from this difeafe, according to Dr Cleghorn, who
examined near one hundred bodies, exhibit conftantly a blacknefs or total corruption of one or o-
ther of the adipofe parts in the lower belly, as the
omentum, mefentery, colon, &c. enlargement of
the fpleen; together with fuch a foftnefs and
rottennefs of its fubftance, that it has more the
appearance of coagulated blood wrapped up in
a membrane, than of an organic part; the gall-
bladder full and turgid; and the ftomach and inteftines overflowing with bilious matter.   No extraordinary appearance was obferved in the cavity of the thorax or head, except the effufion of
yellow ferum where the fkin had been tinged with
that colour.

CASES

* For a more particular account of the fymptoms, by which a
judgement can be formed of the probable event of intermittent fevers, fee Dr. Cleghorn on the Difeafes of Minorca, p. 167, et feq.
from whofe obfervations the above detail is chiefly borrowed.

CASES of Intermittent Fever.

CASE I. (xxv. 4.)

A CLERGYMAN, aged fixty years, who had been troubled for about thirty years with weaknefs of the head and ftomach, thirft, fudden faintings, efpecially when he was in the erect pofture, together with a fenfe of ftraitnefs of his cheft, and with an intermiffion of the pulfe, became at laft affected with a double tertian fever. The fever having grown every day more violent, at laft terminated fatally.

Appearances on Diffection.

ABDOMEN. The omentum was very large, and was twifted like a rope. The right kidney was wanting; and there were no traces of the correfponding renal veffels. The left kidney was of the natural fize, and contained a hydatid full of fluid.

THORAX. In the right ventricle of the heart there was a pretty large polypous concretion, which extended a confiderable way into the vena cava. In the left a concretion of a fmaller fize, that went into the pulmonary vein, was obferved.

HEAD. A confiderable quantity of fluid was found within the ventricles of the brain. The glands on the plexus choroides appeared very much fwelled.

CASE

CASE II. (XLIX. 8.)

A VIRGIN, aged twenty two years, having been affected for many years with a double tertian, and eventually an acute, fever, attended with a pain in the head and the whole body, was thereby carried off.

Appearances on Diſſection.

ABDOMEN. At the extremity of the ileum, in that part which is connected with the mefentery, feveral fmall bodies projected, which in magnitude, form, and colour, refembled the granules of gunpowder. On the internal furface of the uterus feveral round bodies like glands were obferved, on the rupture of which a vifcid humour was difcharged.

THORAX. The lungs were flightly inflamed in thofe parts placed towards the back. The right ventricle of the heart contained a polypous concretion.

CASE III. (LIX. 18.)

A LITTLE boy, affected with tertian fever, became quite emaciated; and at laft, violent convulfions having fupervened, died.

Appearances on Diſſection.

ABDOMEN. The inteftines appeared drawn in
towards

towards the mefentery, which was contracted; and
their coats were fomewhat rigid, and as it were
dried.   They contained, as did alfo the ftomach,
a great quantity of æruginous bile, which ftained
the fcalpel of a violet colour.   Two pigeons being
flightly wounded with a knife dipped in the fame
bile, fo that the bile remained in the wound, in a
fhort time became affected with tremors, and died
convulfed.   A cock, too, who had fwallowed a
piece of bread foaked in that bile, died in a fimilar
manner.

CAUSES of Intermittent Fever.

PREDISPONENT CAUSE.   It has of late become
fafhionable to regard debility as the univerfal pre-
difponing caufe to difeafes; accordingly, it has
been alleged to be that of intermittent fever.   Unlefs, however, the term debility be underftood in a
very unufual fenfe, viz. as being an impaired degree of action of fome of the functions either natural or animal, and not of the whole, it cannot be
deemed predifponent to that difeafe: for it is well
known, that perfons apparently in the full vigour
of health are liable to the attacks of intermittents.
The circumftance predifpofing to the difeafe muft
be that which renders the body fufceptible of the
impreffion of the exciting caufe; it may therefore

2   exift

exift in one organ only, or in feveral, or in all.
The concurrence of many circumftances renders
it probable that, in intermittent fever, the predifpofition takes place principally in the ftomach or
alimentary canal ; but our knowledge of the nature of the proceffes carried on in thofe organs is
at prefent too imperfect, to enable us to fpecify the
particular deviation from nature which conftitutes
or occafions that predifpofition. It is a remarkable fact, that, after intermittents have once occurred in any individual, a ftrong predifpofition to the
return of the difeafe for ever after remains in that
individual.

EXCITING CAUSE. It is now univerfally acknowledged that the moft frequent exciting caufe
at leaft is the effluvia or vapour arifing from
ftagnant water or marfhes, which have been ftyled
Marfh Miafmata. It has been thought neceffary to
afcertain the properties of thofe effluvia, in order
to difcover their mode of action; and accordingly
many ingenious conjectures have been formed on
the fubject. As the effluvia proceeding from great
lakes, falt water, and fogs, do not produce intermittent fevers, it has been fuppofed that thofe
which do fo, confift of fomething more than
fimple watery particles. In marfhes, the putrefaction both of vegetable and of animal matter is
commonly always going forward ; and hence the

moft

moſt general conjecture has been, that vegetable
or animal putrefaction imparted a peculiar quali-
ty to the watery particles of thoſe effluvia. Some
well-atteſted facts, however, clearly overturn this
opinion. Thus, for example, Dr Ruſh aſſerts that
intermittent fevers were totally unknown in ſeveral counties of Penſylvania, until the eſtabliſh-
ment of mill ponds *. Dr Fordyce mentions
too, " that during the war which took place in
Flanders between the tenth and eleventh years
of the preſent century, an army encamped upon ſandy ground, in which water was found in
digging leſs than a foot deep, and occaſioned a
great moiſture in the air; which produced in a
few dáys numbers of fevers, although the army
was perfectly healthy before; and no more fevers
were produced on ſhifting their ground †." It
may therefore be concluded, that we are not yet
acquainted with all the circumſtances which are
required to render marſh miaſmata productive of
intermittents. When it is recollected that intermittents are moſt frequent in warm climates, and
in thoſe climates prevail moſt generally during
warm weather, it is moſt obvious, that the atmoſphere

* Vide, Medical Obſervations and Enquiries, by Benjamin Ruſh,
M.D. vol. ii. p. 265.

† See, A Diſſertation on Simple Fever, by George Fordyce, M.D.
p. 147.

sphere must have a considerable influence in promoting the action of the miasmata; but with respect to the particular nature of that influence nothing more than conjecture can be offered.

It has been imagined, by some respectable authors [*], that intermittent fevers are communicated by contagion. This opinion, however does not coincide with the general observation of practitioners; and if it were just, it must have been long ago established beyond the possibility of contradiction. The author of these remarks has conversed on this subject with several judicious practitioners, who have practised for many years in the countries where intermittents prevailed; and he has been uniformly assured, that in no instance could they ever trace the origin of the disease to contagion.

Many other circumstances have been also regarded, by different authors, as exciting causes of intermittent fever; such as, fear, exposure to cold, watery diet, as that consisting of particular fruits, namely, water melons, cucumbers, &c. and the suppression of habitual evacuations, or recession of cutaneous eruptions. But it is, perhaps, more than probable that these circumstances act only by inducing that state of the body which predisposes to

O 2 the

[*] Dr Cleghorn, p. 132. Dr Fordyce's Observations on Simple Fever, p. 111.

the difeafe ; fince there are many countries where
intermittents never occur, although fome of the
inhabitants muft inevitably be often affected with
fear, and expofed to cold, and have occafionally a
fuppreffion of habitual evacuations, or the receffion of cutaneous eruptions.

Upon the whole, therefore, it may be concluded,
that the fole exciting caufe of intermittent fever
is a ftate of the atmofphere arifing principally from
the evaporation of water, but not proceeding
folely from that circumftance, being influenced by
other caufes not yet afcertained. The late Dr
Cullen regarded marfh miafmata to be the fole exciting caufe. But proof has been adduced above,
that the evaporation from damp ground, and from
mill ponds, produces the fame difeafe ; and it may
be added that, in fome countries, after having been
long prevalent, intermittents difappear entirely * ;
while, in other countries, where they never occurred, they become frequent, although there be no
remarkable change on thofe countries with refpect
to marfhes. Befides, in certain warm climates
where there are no marfhes, intermittent fevers
prevail during the rainy feafon †.

Proximate

* See Dr Currie's Hiftorical Account of the Climate and Difeafes
of the United States of America, page 8.

† See Dr Mofely on Tropical Difeafes, p. 14.

PROXIMATE CAUSE. Thofe who confider all fevers to be effentially of the fame nature, differing only from each other in fymptoms which are influenced by a peculiarity of habit, or of fituation, have afcribed the phenomena of every fpecies to the fame proximate caufe. It is obvious, however, from the obfervations formerly made, that continued fevers are perfectly diftinct from intermittents, in their effential characters, in their progrefs, and in their exciting caufe; and confequently, it is unphilofophical to allege that they depend upon the fame proximate caufe.

Neither lentor of the blood, nor impaired energy of the brain, nor fpafm of the extreme veffels, nor fimple debility, nor an over proportion or accumulation of carbon and hydrogen, can be regarded as the proximate caufe of intermittent fever.

A redundancy of bile in the ftomach and inteftines has been by fome propofed as the proximate caufe of that difeafe; but not only is the bile often accumulated in great quantity in thefe vifcera, but alfo is it even mixed with the blood throughout the whole fanguiferous fyftem, without any fuch effect being produced.

When the phenomena of intermittent fever are carefully confidered, many circumftances appear in favour of the old opinion of Helmont, that they

proceed

proceed from fome affection of the ftomach. The
fhivering, the ficknefs, the thirft, the pain in the
head, all feem to originate from this caufe. The
appearances on diffection, too, in thofe who die of
the difeafe, prove clearly that the fyftem of the
ftomach is very materially affected. But the circumftance which chiefly favours this opinion is,
the great fufceptibility of a renewal of the difeafe
from very flight exciting caufes; as, from the pre-
valence of the eaft wind, and from the repetition
of the original exciting caufe. In this circumftance intermittents differ effentially from all other
fevers: for it is well known, that after continued
fever has once occurred, the perfon fo affected is
by no means fo liable to a return of the difeafe as
one in whom the diforder has never taken place.
In eruptive fevers too, as in the fmall-pox and
meafles, the fever produced by the contagious
matter can never be again produced in the fame
perfon. It may be added, in confirmation of all
this reafoning, that medicines, received into the
ftomach previous to an expected paroxyfm of intermittents, fometimes produce almoft immediate
effects by preventing entirely the fit. But although it were admitted, that the proximate
caufe of the difeafe is a deranged ftate of the ftomach and its appendages, it would ftill remain for
enquiry to afcertain what that deranged ftate is.
                                          Our

Our prefent imperfect knowledge does not enable us to enter on fuch an enquiry.

OBSERVATIONS on the Cases of Intermittent Fever.

The preceding cafes afford very little evidence refpecting either the caufe or effects of intermittents; and afford, too, very little room for remark. In the firft cafe, no morbid appearance was noticed in the ftomach; but a confiderable quantity of fluid appeared within the cranium. The fecond cafe is an example of the intermittent having terminated in continued fever. In the third cafe, the over proportion of bile frequently obferved in that fever was found; but none of the other appearances enumerated by Dr. Cleghorn, and already ftated in thefe pages, were feen.

General Observations on Fevers.

The importance of the fubject induces the author of thefe remarks to deviate from his general plan, by offering a few additional obfervations on fevers.

In the preceding pages, fynocha, typhus, and intermittents, are regarded as difeafes very different

2   in

in their nature. Since, however, an oppofite doc-trine, founded on very plaufible arguments, has been received among practitioners of the firft eminence, it cannot be deemed improper to inveftigate the queftion at fome length.

The fift argument in favour of the opinion, that all fevers are of the fame nature, is, That perfons of different habits of body, expofed at the fame time to the exciting caufe of fever, have fe-vers of a different fpecies. Thus Guido Fanno mentions, that an epidemic prevailed at Leyden, during the Summer of 1669, which affected fome perfons with continued fever, and others with in-termittent *.

In contradiction to this argument, it is to be obferved, that where a number of perfons of oppo-fite temperaments have been expofed to the fame exciting caufe, although fome anomalous fymp-toms appeared in particular inftances, the fevers in confequence have proved effentially of the fame nature †. With refpect to the epidemic at Ley-den, it is not inconfiftent with obfervation to al-lege that two different fpecies of epidemics had prevailed at the fame time; for it is well known that this fometimes happens. Dr. Hillary, for ex-ample,

---

* Vide Haller's Difput. Pathol. p. 204.

† Vide Sir John Pringle's Obfervations on the Jail or Hofpital Fever, § 2.

ample, remarks, that dyfentery and catarrhous fever were, during 1757, epidemic at the fame time in the ifland of Barbadoes *.

The fecond argument in favour of this doctrine is, That intermittents fometimes degenerate into continued fever, and *vice verfa.* Many affertions of this circumftance occur in the writings of authors of the firft eminence †. It muft however be allowed, that the evidence on the fubject is fomewhat ambiguous : for in certain places of Great Britain, as in the fens of Lincolnfhire, &c. intermittents are known to exift for a great length of time without being converted into continued fevers ; and in thofe parts of the Ifland where intermittents do not prevail, continued fevers are never obferved to terminate in intermittents. The argument under confideration, therefore, is not fupported by the phenomena of fevers in Great Britain ; but refts entirely on thofe which appear in warm climates. The evidence in that refpect neverthelefs is not perfectly fatisfactory. Dr. Cleghorn fays, that the tertian fevers epidemic in Minorca, " about the time of the equinox, affume a furprifing variety of forms ; and very often counterfeit continu-

Vol. I.   P   ed

---

* Obfervations on the Epidemic Difeafes in Barbadoes, by Wm, Hillary, M. D. page 125.

† Vide Cullen's Firft Lines, par. xxx. & xxxi.

ed fever, having long redoubled paroxyfns *."
May it not therefore be prefumed, that the inac-
curate obfervation of practitioners is the only foun-
dation for the opinion, that fevers in warm climates fpontaneoufly vary their type during their
progrefs?   Befides, although it were proved beyond the poffibility of doubt, that intermittents
terminate in continued fever, and *vice verfa;* no
evidence is thereby afforded that the difeafes are
of the fame nature : for, by fimilar evidence, dyfentery and intermittents fhould alfo be of the
fame nature, fince they too, under certain circumftances, are apt to terminate in each other †.

But the principal circumftance by which the
doctrine alluded to has acquired influence, is the
commonly received opinion, that fevers are produced by a number of exciting caufes of an oppofite nature.   If contagion, human effluvia, marfh
miafmata, expofure to cold, fatigue, intemperance,
and paffions of the mind, all feverally induce the
fame difeafe ; it muft follow, that the varieties in
the fymptoms obferved, in particular inftances, depend upon peculiarity of conftitution, and not
upon the nature of the difeafe.

Although, in the preceding pages, the author
of

* Cleghorn, loco citato, page 185.
† Vide Cleghorn's Obfervations, p. 134 ; and Dr. Mofely on
Tropical Difeafes, p. 121.

of thefe remarks has alleged, that expofure to
cold, paffions of the mind, intemperance, &c.
where they do excite fever, produce only fynocha; that fpecific contagion, a peculiar ftate of
vegetable putrefaction, and human effluvia, are
the fole exciting caufes of typhus; and that marfh
miafmata, and the vapour arifing from ftagnant
water, under certain circumftances, are the only
exciting caufes of intermittents; he is well aware
that feveral objections may be urged againft the
opinion. Some of thefe have been already noticed, and, it is thought, repelled. One however
ftill remains to be confidered; viz. That during
the convalefcence fucceeding typhus, expofure to
cold, intemperance, paffions of the mind, &c. are
frequently obferved to occafion what is termed a
Relapfe, or a return of the difeafe, which had
difappeared. Any reafoning upon this fact muft
be avoided in this work; as it would lead to dif-
cuffions by far too extenfive. The circumftance
probably depends upon that law of the animal
oeconomy, by which recently deranged functions
are very readily again affected.

SECT. III. *HECTIC FEVER.*

HECTIC FEVER, being fymptomatic of fome previous difeafe, never invades fuddenly; but attacks the patient by flow, and in general infidious, approaches *. In this difeafe there are two exacerbations every twenty-four hours; in which refpect in coincides with remittent fever. The firft of thefe exacerbations occurs about noon; and the fecond towards the evening. Each of them is ufhered in by a chilly fit; not always diftinctly marked, as it appears often to be nothing more than an increafed fufceptibility of the impreffion of cold. A dry burning heat foon fucceeds; attended with a florid red colour in the cheeks. This colour however is confined within a circumfcribed fpot. The pulfe is frequent in the number of pulfations, and quick in the contraction of the artery; and at that time pretty full. The tongue is clean, as it commonly is through

the

---

* For an accurate defcription of hectic fever, fee Dr. Cullen's Firft Lines, par. 858, et feq.; and Medical Tranfactions, vol. ii. page 1.

the whole of the difeafe. Some degree of thirft is
felt. There is feldom any headach ; and never,
till towards the fatal termination of the complaint,
any delirium. The urine is high coloured ; de-
polites a red branny fediment ; and is fometimes
covered with a fatty film. The patient feels a
certain uneafinefs ; more remarkably obferved during the evening exacerbation, as it induces reftlefsnefs and watching ; but he is unable to afcertain the circumftance from which his uneafinefs
proceeds. By degrees the remiffion takes place ;
and he thinks himfelf free from every complaint. He continues however to have a hot dry
fkin, a quick weak pulfe, a particular palenefs in
the face, a pearly whitenefs in the eyes, and confiderable proftration of ftrength. The appetite for
food is often not much impaired. The mind, with
very few exceptions, is chearful ; no apprehenfion
of danger being entertained. The belly, at the
beginning, is commonly bound. As the difeafe
advances, many of thefe fymptoms are aggravated,
and fome additional ones occur. Thus the fhiverings come on frequently during the hot fit. The
evening exacerbations terminate in profufe colliquative fweats. Great emaciation of the body takes
place ; fo that the hairs fall off, and the nails of the
fingers become adunque. The eyes are hollow ; the
belly is drawn inwards ; and the fkin grows hard.

Diarrhœa,

Diarrhœa, frequently alternating with the morning fweats, fuperyenes. The feet are affected with œdematous fwelling. The tongue and fauces appear inflamed, and at laft are covered with aphthæ. Soon after which death, preceded often for fome days by delirium, enfues.

The progrefs of hectic fever is fo flow, that the death of the patient may be generally with certainty prognofticated many weeks before it happen.

The feat of the difeafe is evidently the vafcular fyftem.

Hectic fever, as already mentioned, is fymptomatic of fome other diforder. The difeafes of which it is a fymptom, to be noticed in this work, are; *Firft*, Phthifis Pulmonalis; *Secondly*, Tabes Mefenterica; and, *Thirdly*, Certain anomalous cafes of tabes.

─────────

§ 1.   PHTHISIS PULMONALIS.

EMACIATION and weaknefs of the body, attended with cough, hectic fever, and moft generally the expectoration of purulent matter *, is the

beft

* Wide Cullen Nofolog. Ed. 1785. p. 397

beſt definition of phthiſis pulmonalis that can be given.

As this diſeaſe ariſes from exciting cauſes of op-poſite natures, and is attended in different caſes by different ſymptoms, authors have divided it in-to ſpecies.   Some have founded their arrangement on the ſymptoms * ; others, on the remote cau-ſes †; and ſome on the ſtages of the diſeaſe ‡.

The author of theſe remarks propoſes, with much diffidence, an arrangement derived from the diſ-eaſed ſtate of the lungs.   In doing this he offers two reaſons ; *Firſt*, That he apprehends the ſymp-toms of the diſeaſe to be different, according to the morbid affection of the lungs ; and, *Secondly*, That he believes ſuch an arrangement to be ſubſervient to practical purpoſes.

Phthiſis pulmonalis may therefore, in his opi-nion, be divided into three ſpecies ; viz. Phthiſis a Tuberculis, Phthiſis a Materia calculoſa, and Phthiſis a Vomica.

The following general obſervations are appli-cable to all the ſpecies of the diſeaſe.

Although

* Vide Macbride's Works, quarto edition, p. 397.

† Vide Richardi Morton Opera, p. 109. and Dr White's Obſerva-tions on Phthiſis Pulmonalis, publiſhed by Dr Hunter of York, p. 94.

‡ Vide Cullen Noſolog. Edin. ut ſupra, p. 159 ; and Ruſh's Medical Enquiries, vol. ii. p. 111.

Although the progrefs of Phthifis Pulmonalis be commonly flow, its termination, in by far the greater number of cafes, is uniformly fatal; and indeed it has been queftioned whether recovery from the difeafe ever took place. The fatal termination, however, is fometimes protracted for a confiderable length of time.

It is remarkable that apprehenfion of danger is feldom entertained by the patient himfelf; for it often happens, that he finks under the difeafe without being in the leaft fenfible of his approaching diffolution. But, in thofe rare cafes, where the living principle feems to be extinguifhed with great difficulty, where aphthous ulcerations appear in the fauces, and where, from confinement to bed for a confiderable length of time having become indifpenfible, gangrene takes place in thofe parts of the body which fuffer the greateft preffure, the patient feems fenfible of his hopelefs condition for a fhort time before death.

It is a curious circumftance, that this difeafe is fufpended by pregnancy, and is fometimes cured by mania. It has been generally imagined that the life of a patient affected with phthifis pulmonalis is protracted by pregnancy; but there is reafon to believe the opinion to be erroneous. In order to judge of this, the ordinary progrefs of the difeafe, where it is uninterrupted, ought to be compared with that where it is fufpended by pregnancy. If this be

2   done,

done, it will be generally found, that, if two women be affected with phthifis at the fame time,
and one of them fhall become pregnant, the other
fhall live confiderably longer, although in her
the difeafe proceeds uninterruptedly through its
feveral ftages. This circumftance has occurred
repeatedly to the obfervation of the author of thefe
remarks; who has uniformly found, that women
labouring under phthifis, previous to pregnancy,
fink within two or three weeks after delivery.
Experience has alfo taught him, that the difeafe is
not always fufpended by pregnancy; for he has
feen feveral cafes where fuch women have died
during the fixth or feventh month of geftation,
fometimes undelivered, but moft frequently with
previous abortion *.

Phthifis generally attacks at fome period between the age of puberty and the thirty-fifth
year.

VOL. I.   Q   PHTHISIS

---

* The author fome years ago attended a patient whofe cafe was
very fingular. Incipient phthifis was fufpended by pregnancy, and
apparently cured by mania having fupervened to delivery. For
fome months after the mania had ceafed, the patient continued in
perfect health; but fymptoms of phthifis then recurred, and at the
fame time fhe became pregnant. The difeafe proceeded fo rapidly, that premature labour took place about the fixth month of geftation; fhe was delivered of a ftill-born child; and expired within
twenty four hours.

PHTHISIS A TUBERCULIS. This is by far the most frequent species of the disease. It occurs only in those of a scrophulous habit, and in those born of scrophulous or of phthisical parents. It is commonly preceded by the following symptoms, enumerated by Dr Rush *; the principal of which were first noticed and accurately detailed by Morton † :—" A slight fever, increased by the least exercise ; a burning and dryness in the palms of the hands, more especially towards the evening ; rheumy eyes upon waking from sleep ; an increase of urine ; a dryness of the skin, more especially of the feet, in the morning; an occasional flushing in one and sometimes in both cheeks ; a hoarseness ; a slight or acute pain in the breast ; a fixed pain in the side ; a shooting pain in both sides; headach ; occasional sick and fainty fits ; a deficiency of appetite ; and a general indisposition to exercise or motion of any kind." Sometimes, however, these symptoms do not take place, or are not observed.

The first symptom of this species of the disease is cough, which at the beginning is generally slight, and unaccompanied with expectoration. After the cough has continued for a considerable length of time, being sometimes more and sometimes less

violent,

---

* Rush's Medical Enquiries, vol. ii. p. 107.
† Vide Richardi Morton Opera, p. 39.

violent, and being aggravated by the moſt apparently trifling cauſes; hoarſeneſs of the voice, a
ſenſe of weight and ſtraitneſs in the cheſt, felt
more eſpecially after motion; inability to lie on
one or both ſides, and ſlight difficulty of breathing ſupervene. The pulſe from the firſt is frequent, and for a conſiderable time full and hard.
The breathing is much quicker than that of a per-
ſon in health. It is performed with great motion
of the cheſt, and is frequently accompanied with a
noiſe ſimilar to that of ſighing. The cough gradually becomes more violent; it attacks by fits,
which are particularly diſtreſſing towards the
evening and during the night; and the expectoration of mucus, and afterwards of pus, takes place.
Theſe ſymptoms are attended with occaſional fly-
ing ſtitches in the breaſt and ſides, great laſſitude,
emaciation of the body, and inequality of temper.
The appetite for food is diminiſhed, and vomiting
frequently occurs after eating. Blood is generally ſpit up, ſometimes in ſmall and ſometimes in
large quantities; often before any appearance of
matter can be perceived, and not unfrequently a-
long with pus. The expectoration, therefore, at
this period of the diſeaſe, is various in its appearance; ſometimes reſembling mere mucus; at
other times, pus, blood, or a mixture of both;
ſo that it is of very various colour and conſiſtence;
being at times white, grey, brown, yellow, red,

Q 2   and

and green, and fometimes ftreaked with all thefe
colours; and being at one time thick, at another
time thin, and in fome cafes being tough, and imi-
tating the figure of the branches of the bronchia.
It is often difficult, at the beginning, to afcertain
whether the matter expectorated be mucus or pus.
Many experiments have been made, with the view
of eftablifhing a criterion between the two. This
however, ftill remains a defideratum, probably for
the reafon mentioned by Dr. Stark, viz. that the
fpitting of confumptive perfons combines the effential properties of mucus and of pus*.

To the fymptoms above enumerated, hectic fever is added. The progrefs of the difeafe, after
this period, is, comparatively fpeaking, rapid, as it
generally terminates fatally within a few months
from that time.

Authors have divided the fymptoms of this fpecies of phthifis into two ftages; the firft of which
they ftyle Inflammatory, comprehending all the
fymptoms previous to the occurrence of hectic fever; and the fecond they term Putrid or Suppu-
rative.

The appearances on diffection exhibit roundifh,
white, firm bodies, of different fizes, from the
fmalleft granule to about half an inch in diameter,
but moft ordinarily of the fize of a garden pea.

named

---

* Vide The Works of the late Dr. Stark, p. 23.

named *Tubercles*, fituated in the cellular ftructure which connects the air cells of the lungs together. The fmall tubercles are frequently accumulated in clufters. When cut into, they appear to confift of a folid white fubftance, almoft as firm as cartilage, having the cut furface, fmooth, fhining, and uniform, and being covered with a thin capfule. The larger tubercles confift alfo of a curdly kind of pus; and the fame matter is obferved in cutting into the clufters of the fmaller tubercles; but its origin cannot be afcertained. The cavity of the large tubercles always communicates with branches of the trachea. Neither cells, veficles, nor veffels have been difcovered in tubercles. The branches of the trachea communicate alfo with cavities of a larger fize (the largeft however not exceeding four inches in extent), called *Vomicæ*. Thefe cavities contain pus of various appearances and confiftence in different cafes. With this they are not filled, as they have the furface only befmeared with it. They are lined with a fine delicate membrane, fimilar to the capfule of the tubercles. They communicate with branches of the trachea by round fmooth openings, and with one another by ragged and unequal ones. They are generally fituated towards the back part of either lobe, and are commonly concealed. Many circumftances concur to fhew that the vomicæ are modifications of the tubercles. The lungs contiguous to the tubercles

and

and vomicæ are fometimes in a natural ſtate ;
more often indurated. The pulmonary arteries
and veins, as they approach the larger vomicæ, are
ſuddenly contracted and obſtructed, ſo that they
have little or no communication with thoſe cavities. This contraction does not take place in the
vicinity of vomicæ under an inch in diameter. In
ſome caſes, theſe morbid changes are very exten-
ſive; and in other caſes, they are limited to particular parts of the lungs only. In the latter in-
ſtance, the ſuperior and poſterior parts of thoſe organs are alone diſeaſed; and in the former, the
ſame parts are principally affected. The lungs of
the left ſide are more commonly diſeaſed than thoſe
of the right *.

PHTHISIS A MATERIA CALCULOSA. This ſpecies
is characteriſed by a ſhort cough, unattended with
expectoration; ſhort, frequent, and difficult breathing, which is neither relieved nor aggravated by
poſture ; pain in the thorax ; and ſometimes excef-
ſive diſcharge of blood from the lungs. It is peculiar to perſons above thirty years of age ; while
the phthiſis a tuberculis occurs moſt frequently before that period. The progreſs of this ſpecies is
                                                    not

* The above deſcription of the appearances on diſſection is borrowed from Stark's Works, p. 26, et ſeq.; and Bailie's Morbid A-
natomy, p. 46.

not fo rapid as that of the former. The patient generally dies during a fit of breathlefsnefs.

The appearances on diffection exhibit earthy concretions, formed throughout the fubftance of the lungs; and feated either in the extreme branches of the bronchia, or in the cells connected therewith.

PHTHISIS A VOMICA. This fpecies is generally preceded by pneumonia, or fome accidental injury of the lungs. The expectoration, from the beginning, is purulent, and in confiderable quantity; and the cough is trifling. The breathing is commonly frequent and fhort; and inability to lie on one or other fide is experienced. The progrefs of the difeafe is commonly very rapid.

Appearances on diffection fhow extenfive abfceffes in the lungs, independent of tubercles. Thefe abfceffes, called Vomicæ, (as well as thofe connected with tubercles) are feated fometimes within the fubftance of the lungs, and fometimes on the external furface; in which cafe an adhefion is generally found between the affected part and the pleura. The death of the patient, in this fpecies, is commonly very fudden; and is immediately occafioned by the burfting of the abfcefs.

CASES

CASES of Phthisis Pulmonalis.

Phthisis a Tuberculis.

C a s e I. (xxii. 15.)

A strumpet, aged about twenty years, had been affected for many years with a flow fever, a cough, an ill conditioned expectoration, and emaciation of the whole body. She complained of pain in the left fide of the thorax; and when fhe lay on that fide was troubled with difficulty of breathing. To thefe fymptoms a copious fpitting of blood fupervened. This was checked: but two days after, while a violent fouth wind prevailed, in which ftate of the air thofe who are affected with a diforder of that kind generally die, fhe expired.

Appearances on Diffection.

Thorax. The right lobe of the lungs adhered very flightly to the ribs. In the fubftance of both lobes a number of hard tubercles, of a whitifh colour, and refembling glandular bodies, were obferved. The fuperior part of both lobes was alfo affected with other diforders. For in the right lobe, towards the fternum, a large hollow ulcer, containing purulent matter, was feen; and the left, towards the fide, contained a hard fubftance,

3   equal

equal in fize to a large pear, refembling in fome meafure the fubftance of the pancreas when indurated.   In the middle of this body there was a fmall ulcer full of pus.   In the pericardium there was little ferum.   In the left ventricle of the heart a fmall polypous concretion was found; and the right contained a polypus of a moderate fize, the greater part of which was inferted into the neighbouring auricle.

Case II. (xxii. 14.)

A musician, middle aged, who had three years before been affected with fpitting of blood, had afterwards been troubled with cough, attended with the expectoration of the matter ftyled defluxion. After many months had elapfed, the fpitting of blood returned; and afterwards it again took place, and left behind it the expectoration of a great quantity of thick matter of a very bad appearance. He could lie eafily on either fide; and had no pain in the thorax.   His cough however was troublefome during the night, and efpecially after fupper.   His breathing was difficult, particularly after motion, even of the flighteft kind.   Along with thefe fymptoms, he had great thirft and oppreffion at the ftomach after taking food.   For fome weeks before his death he had frequent nocturnal fweats.   Although it had been ufual for

Vol. I.   R   his

his feet to be fometimes fwelled, and afterwards
to fubfide entirely; yet during the latter days of
his exiftence they no longer fubfided.   Diarrhœa,
by which much ferous matter was difcharged,
having fupervened, he died as he was beginning
to drefs himfelf in order to rife.

Appearances on Diffection.

THORAX.   The lungs were filled with many
tubercles.   The fuperior lobule of the left lobe,
at the upper part towards the fternum, was externally very much indurated; and contained internally a pretty large ulcer, in which fanious matter like a poultice was found.   In the right cavity of the thorax almoft half a pound of ferous fluid
was feen; and in the pericardium there was the
fame quantity.   The ferum in that latter part
having been expofed to the fire, completely difappeared, leaving only a pellicle at the bottom
of the veffel.

CASE III.   (XXII. 18.)

AN unmarried woman became affected with a
fever, faid to be in confequence of a fright.   The
fever was attended with a pain in her breaft.   The
parotides, and almoft all the glands of the neck,
were fwelled.   She died.

Appearances on Diffection.

ABDOMEN.   The abdomen contained a fmall
quantity

quantity of limpid watery fluid. The omentum was connected with the mesentery and peritoneum by a kind of small ligaments. The surfaces of the peritoneum, the omentum, the mesentery, the intestines, the uterus, the gall-bladder, and the urinary bladder, were unequal; in consequence of protuberating bodies which lay here and there at a distance from each other, and were of various shapes and sizes. In the upper part of the omentum they were exceedingly small; in the inferior part they were very large and numerous, and lay quite contiguous to each other.

THORAX. In the left lobe of the lungs, not only was there an ulcer containing sanious ichor, but also substances, similar to those which were found in the mesentery and in the other parts of the belly, appeared. Some of these bodies contained pus; some a matter almost like a poultice in its consistence; while some were so solid, that they resembled natural conglobate glands.

CASE IV. (XXII. 24.)

A PHYSICIAN, middle aged, who had long had a cachectic appearance in his face, and had for some time been affected with difficulty of breathing and hoarseness of voice, at length began to spit up a variously coloured matter. Along with this matter he one day coughed up a small curved

R 2   bone,

bone, (but not very minute) which was fmooth
on the concave part, and rough on the convex.
After this he was pretty frequently affected with
a fenfe of fuffocation.   At length one night, after
having faid, before he went to bed, that he felt
himfelf better, he was found dead.   His death
took place without being perceived by a perfon
who lay by him for the purpofe of taking care of
him.

Appearances on Diffection.

THORAX.   In the lungs, both externally and
internally, a number of veficles, filled with white
pus, of various fizes, the largeft not exceeding
that of a grape, was obferved.   The pericardium
contained a confiderable quantity of turbid fluid.
In the heart one very fmall polypous concretion
was found.   The other parts within the thorax
could not be examined leifurely, as the body was
opened without permiffion from the relations.

C A S E V. (xxvi. 29.)

A WOMAN, who had been for a long time thought
to be confumptive, was found to have died fuddenly.

Appearances on Diffection.

THORAX.   In the left cavity of the thorax the
correfponding lobe of the lungs contained three
or four tubercles full of pus.   In the right cavity
the

the lungs were found ; but coagulated blood, to the quantity of four pounds, was found extravafated. This blood had been effufed from the trunk of the vena azygos. That veffel, although it had collapfed in confequence of the effufion, was ftill fo large, that it might be compared to the vena cava. The dilated portion was about the length of a fpan, and in its middle, an open foramen of the form of an elipfis was perceived. Through this foramen the extravafated blood had paffed.

PHTHISIS A MATERIA CALCULOSA.

CASE I. (XV. 25.

A WOMAN affected with a very flight fcabies, and confiderable wafting of flefh, was troubled with a fhort cough, which was never attended with a difcharge of thick expectoration, but always with a difficulty of breathing. This latter complaint was neither increafed nor diminifhed by pofture, as it continued the fame whether fhe lay on her back or on either fide. When her neck was raifed, indeed, fhe breathed a little more eafily ; but then fhe felt the fenfation of a weight extending from the fauces into the cavity of the thorax, and rendering the fauces narrow. With thefe fymptoms fhe died.

Appearances

Appearances on Diffection.

THORAX. The lungs were as it were tartariza-
ted; for wherever they were cut into, the knife
made the fame rafping kind of noife as if one had
cut into fandy.concretions.

C A S E II. (LXVIII. 12.)

A YOUNG gentleman, aged fifteen years, of a
good complexion, who had been healthy from his
infancy, having been very feverely chided, and, as
he was naturally thoughtful, having been much affected thereby, continued for the fucceeding three
days almoft in a ftate of ftupor. On the third
month after this, he begun to obferve fmall glands
on his neck, which at firft increafed in fize gradually; but, in a fhort time the increafe of the
difeafe became fo rapid, that befides the fwelling of
the falivary and axillary glands, prominent bodies
of the fize of a pigeon's egg, refembling ftrumous
glands, were felt on the back, the breaft, and more
efpecially about the clavicles. Tumours of the
fame kind were foon after felt on the integuments,
and alfo within the cavity of the abdomen. All
thefe tumours were indolent, except one lying on
the pectoral mufcle, which was three inches in
length, livid in its colour, and fomewhat painful
to the touch. Thofe within the belly, too, particularly on the left fide, where there was a great
degree

degree of tenfion, along with refiftance when preffed, were painful. Not long after the beginning
of the difeafe, an acute pain was felt fometimes in
the right knee and leg, and fometimes in the left.
From that period he was always unwell, being
affected with fymptomatic and irregular feverifh
attacks, with watchings, and with progreffive emaciation. He was neverthelefs lively and cheerful
to the very laft. His appetite for food never diminifhed, but on the contrary was very keen, efpecially in the latter days of his life. Although he felt
a certain uneafinefs about his throat, proceeding
from mucus, which was readily hawked up by a
flight cough, he never had any difficulty of breathing; which, confidering the appearances after
death, is very furprifing. Many external and internal remedies were employed from the very beginning, and more efpecially during the progrefs
of the difeafe, when the moft active medicines
were prefcribed by the moft experienced practitioners. But, notwithftanding every means, the
bulk of the tumours increafed, and the difeafe was
fo rapid in its courfe, that, although it had only
commenced in December, the death of the patient
took place towards the end of May.

Appearances on Diffection.

EXTERNALLY. The integuments of the neck,
breaft, and abdomen, being cut into, the external
tumours were found to be feated in the adipofe
membrane,

membrane, which in thofe places was clofely connected by the under-lying mufcles. All thefe tumours were filled with a whitifh matter, partly of
a fluid, but chiefly of a folid and febaceous confiftence. Some of the moft prominent, as thofe adhering to the pectoral mufcles, which formed with
the axillary glands one continued body, difcharged
when cut into a yellowifh and fanious fluid.

ABDOMEN. The whole omentum appeared to
be befet here and there with fmall hard bodies,
filled with the whitifh matter above mentioned.
The liver, the fpleen, and the kidneys were themfelves in a natural ftate; but the following appearances about each of thefe organs were obferved: The peritoneum was ftrongly connected to
the right fide of the liver. It, then, after having
in feveral places formed as it were one and the fame
fubftance with the contiguous mufcles, projected
outwards in the form of a body of the fize of a
hen's egg, filled with the matter already mentioned. This body was connected with the urinary
bladder. Near the fpleen, the left part of the mefocolon formed a fwelling, which confifted of a
congeries of tumours refembling pigeons eggs.
Thefe tumours were filled with the fame matter
obferved in the others. The pancreas was full of
the fame tumours, and fimilar fwellings were feen
fcattered up and down through the mefentery.
The adipofe membrane of the kidneys was mon-

I   ftroufly

ftroufly thickened, being on the part placed towards the vertebræ two inches, and on the oppofite part five inches thick. It was every where
diftended with the matter fo often mentioned.
The left kidney, together with a hard line which
interfected the matter, weighed thirty-fix ounces.
The inteftines were by no means free from difeafe :
For fmall bodies of the fame nature with thofe already feen, were obferved on the adipofe appendages of
the colon, and on the ligamentous bands which pafs
through that inteftine.   The glands of Peyerus, in
the fmall inteftines, in fome places, were of the figure and fize of a lupin.   One of thefe glands,
much larger than the reft, was inflamed, and contained putrid fluid.

THORAX.   The mediaftinum was in the middle
thickly befet with the bodies above mentioned,
and one of thefe of the fize of a fmall hen's egg,
not only lay contiguous to, but alfo compreffed
the trunk of the afpera arteria.   The lungs were
internally found, but on their external furface,
a number of hard and ftony globules of the fize
of barley-corns, were implanted.   The fame furface was on the back part, hollowed out on both
fides, but particularly on the left, in confequence
of tumors of a large fize which had formed on the
adjoining pleura.   Some of thefe tumors were fituated near the dorfal vertebræ ; and others were
fo difpofed that one lay on each rib regularly

VOL. I.   S   from

from the lower part to the upper. The heart was in a natural ftate, except that the external furface of the right auricle was completely granulated, as it were from little bodies of the fame kind.

CASE III. (XLVIII. 38.).

An old woman had her right leg bitten by a dog. This was fucceeded by great thirft, inteftinal flux, and flight fever. After feveral days the thirft abated ; but the other fymptoms continued. She was then affected with vomiting, by which fome worms of the lumbricus kind were thrown up. The vomiting finally ceafing, fhe became more and more exhaufted, and expired. In this patient, the pulfe had never been ftrong, but had fometimes been liable to intermiffions. The woman had alfo been fubject to cough, which however was flight.

Appearances on Diffection.

Abdomen. The ftomach appeared fomewhat diftended with air; and being naturally large, it extended fo low that the portion of the colon which lies next it, was fituated below the navel. The whole of that inteftine, except its beginning, which together with the cæcum was turgid with air, was fo much contracted that it refembled one of the fmall inteftines. The duodenum, on the

contrary,

contrary, was much larger than ufual, and paffed downwards on the right fide over a very long tract of vertebræ. The other fmall inteftines were of a dirty and colour. The mefenteric *fluvid* glands of a moderate fize, were diftinctly perceived under a fmall quantity of fat. The liver was large. Two furrows, as if made by a deep impreffion of the fingers were perceived on the fuperior part of its convex furface, from whence they defcended in a parallel direction for a confiderable length on the fore part. The fpleen was thicker than ufual; it was fomewhat rough on the gibbous furface, in confequence of certain granules as it were; internally it was of a pale colour. The uterus lay towards the left fide. Its whole internal furface was rough, but was not ulcerated; although that of the fundus, but not that of the cervix, was covered with black blood. The parieties of the os tincæ were fomewhat thickened.

THORAX. The lungs were diftended with air. Many of the bronchial glands about and within them were enlarged, and contained a tartareous matter. Both ventricles of the heart were filled with polypous concretions of a yellow white colour, placed amidft fome black blood. Some of thefe concretions were thick, and not eafily torn. The valvulæ mitrales at the lower part, and efpecially at that part next the aorta, were compofed

of

of a compact fubftance, which was internally of a white colour. None of the valves of that artery were perfectly free from incipient offifications. One in particular on the furface, which was turned towards the fide of the artery was completely bony, rough, and unequal, in confequence of particles really offeous, in fome parts lying upon each other, that projected like grains of fand. On the other furface, this valve degenerated into a flefhy excrefcence of a larger fize than itfelf.

HEAD. Air bubbles were obferved in the veffels of the pia mater. A fmall quantity of fluid was found under the fame membrane and within the ventricles. The choroid plexufes were not pale.

PHTHISIS A VOMICA.

CASE I. (XXII. 16.)

AN unmarried woman, aged twenty-four years, after having been affected with fpitting of blood, was troubled with a cough. She expectorated a catarrhous matter, which at laft refembled fanies. She was feverifh, and complained of pain in the thorax, efpecially in the left fide, on which fhe could not lie. Her whole body was emaciated, except her feet which were both confiderably fwelled.

The

The right foot was affected with eryfipelas for fome days before death.

Appearances on Diffection.

THORAX. The left cavity of the thorax was filled with a ferous fluid, together with fmall portions of coagulated blood, adhering in different places to the pleura, and to the lower edge of the lungs. The pleura was as red as if it were inflamed, and the contained portion of the lungs had become indurated as inflamed lungs generally are. The right cavity of the thorax contained little ferum, and in it the pleura was found; but the lobe at that part next the clavicle was fomewhat hard, and concealed an ulcer in the middle of the hardened portion. Scarcely any veftige of fluid could be traced in the pericardium. The ventricles of the heart were completely filled with coagulated blood.

CAUSES of PHTHISIS PULMONALIS.

PREDISPONENT CAUSE. Perfons of a fanguine, or of a fanguineo-melancholic temperament, as it is called, who have a very fine fkin, with the veins fhining through it, a rofy complexion, foft flefh, thick upper lip, a long neck, narrow cheft, prominent fhoulders, and are altogether of a delicate make, and of much fenfibility and irritability;

thofe

thofe the growth of whofe body had been at the
age of puberty remarkably rapid ; thofe who during the early periods of their life had been affected with fcrophula ; and thofe born of fcrophulous
and of phthifical parents, are chiefly predifpofed
to the firft and third fpecies of phthifis.

It appears, therefore, that a particular conftruc-
tion of the body, or of the fyftem, is the predifponent caufe of the difeafe in thofe fpecies. The
ftate of the fyftem on which the fecond fpecies
depends, is exceedingly obfcure. It is probably
fomewhat analogous to that from whence preternatural offification proceeds.

EXCITING CAUSES. Hæmoptyfis, Catarrh, Afthma, Pneumonia; Wounds and Injuries from external violence affecting the Lungs; the Introduction of Extraneous Matters into thofe Organs;
Tubercles of the Lungs; Contagion; Excefs of
Oxygen in the Blood; General Debility; the Repulfion of Cutaneous Eruptions; and the Metafta-
fis of the matter of Syphilis and Scurvy, have been
regarded as the exciting caufes of the firft and third
fpecies of phthifis pulmonalis.

Hæmoptyfis, Catarrh, Afthma, and *Pneumonia,*
are only productive of phthifis pulmonalis where
there is a predifpofition to that difeafe. The fame
obfervations apply to wounds and external injuries
communicated to the lungs : for Dr. Rufh mentions,

tions, that out of twenty-four British soldiers admitted into the hofpitals, during the campaign of 1776, with wounds in their lungs, twenty-three recovered [*].

The Introduction of Extraneous Matters into the Lungs. This happens in confequence of certain occupations in life; as, the grinding of corn, flax-dreffing, &c. But it is not probable that under fuch circumftances phthifis pulmonalis is produced, unlefs there be a ftrong predifpofition to the difeafe. Thus, millers are fubject to a dry cough, which often continues for many years without terminating in phthifis.

Tubercles in the Lungs. The nature of thefe tubercles has been varioufly defcribed by different authors. By fome they have been regarded as enlarged glands [†]; and by others, as effufions of mucus [‡]. From the obfervations of Dr. Stark and of Dr. Baillie [§], it appears, that tubercles are feated in the cellular ftructure which connects the air cells of the lungs together; and confequently, it is inferred by Dr. Baillie, they cannot be glandular, as there are no glands in that membrane. Their circumfcribed form, and their ftructure,

feem

* Vide Dr. Rufh's Medical Enquires, vol. ii. p. 97.

† Vide Dr. Cullen's Firft Lines, par. 876; and Dr. M'Bride's Works, quarto edition, p. 397.

‡ Vide Dr. Rufh's Enquiries, vol. ii. p. 99.

§ See Baillie's Morbid Anatomy, p. 46.

feem to overturn the opinion of their being effufions of mucus. It has been imagined that tubercles are inorganifed bodies; but the following circumftances contradict this. *Firft*, The progreffive increafe in their fize, which is clearly marked; *Secondly*, The converfion of their contents into pus; and *Thirdly*, Their being covered with a thin capfule. Their precife nature ftill therefore remains to be determined. The fuppofition of their being indurated glands was certainly the moft natural one; feeing that they are generally accompanied with glandular fwellings in other parts of the body, and that they occur only in perfons born of fcrophulous and phthifical parents.

Contagion. It is ftill difputed among practitioners whether phthifis pulmonalis be ever communicated by contagion. That it is not a frequent caufe of the difeafe, is evinced by daily experience: but it might be rafh to affert that it can never produce it, efpecially as there are fome ftriking facts adduced to prove that it fometimes acts as an exciting caufe· Thus Dr. Rufh relates, that " the late Dr. Beardfley of Connecticut informed him, that he had known feveral black flaves affected by a confumption, which had previoufly fwept away feveral of the white members of the family to which they belonged *."

Excefs

* See Dr. Rufh's Medical Enquiries, vol. ii. p. 101.

Excefs of Oxygen in the Blood. Dr. Beddoes *
has endeavoured to fhew, that in certain cafes
of phthifis the blood is hyper-oxygenated. This
over proportion of oxygen in the blood, however, though proved, muft necefïarily proceed from
fome deranged ftructure of the lungs, or of the
fyftem. In that cafe it fhould be regarded as the effect of the exciting caufe, and not as the exciting
caufe itfelf: a diftinction necefïary not only for
accuracy of ideas, but alfo probably for practical
purpofes †.

General Debility. Dr. Rufh has alleged, that
general debility of the whole fyftem is the exciting caufe of the firft fpecies of phthifis pulmonalis ; and that fome of the circumftances generally enumerated as fuch, as hæmoptyfis, tubercles, &c. are the effects, and not the caufes, of
the difeafe ; while others, as violent paffions of the
mind, exceffive evacuations, cold and damp air,
&c. act only by producing debility. " Should it
be afked," he obferves, " Why does general debi-
Vol. I.   T   lity

* Obfervations on the Nature and Cure of Calculus, Sea Scurvy, Catarrh, Fever, &c. by Thomas Beddoes, M. D. page 136,
et feq.

† Dr. Beddoes has propofed two theories refpecting the caufe
of confumption. By the one of which, he confiders the difeafed
ftate of the lungs to be the effect of the increafed quantity of oxygen combined with the blood ; and it is that theory which is confidered here. The fecond theory is noticed in another place.

lity terminate by a diforder in the lungs rather than any other part of the body? I anfwer, That it feems to be a law of the fyftem, that general debility fhould always produce as a fymptom fome local difeafe *." He farther obferves, " It would feem as if the debility in the cafes of confumption is feated chiefly in the blood-veffels; while that debility which terminates in difeafes of the ftomach and bowels is confined chiefly to the nerves; and that the local affections of the brain arife from a debility invading alike the nervous and arterial fyftems." It is not eafy to comprehend, notwithftanding the obfervations of the ingenious author of this opinion, how debility fhould produce an affection of the lungs without the intervention of fome exciting caufe. The fymptoms of general debility which he enumerates, as preceding phthifis pulmonalis, viz. " quick pulfe, efpecially towards evening, heat and burning in the palms of the hands, faintnefs, headach, ficknefs at ftomach, and occafional diarrhœa," certainly do fometimes take place before any complaint in the breaft; but in by far the greateft number of cafes, occur only after the affection of the lungs has commenced.

It is well known that phthifis fometimes follows the repulfion of cutaneous eruptions, as fmall pox, measles

---

* Vide Rufh, loco citato, page 105.

measles, &c. and it has been all·ged that it also follows in some cases syphilis and scurvy. This has been accounted for in different ways. Thus Dr Cullen supposes that those diseases produce an acrimonious state of the fluids, which occasions tubercles; and Monf. Portal, as well as Dr Ryan, think that they produce a local inflammatory affection of the lungs, which terminates in ulcerations without tubercles.

There is no inconsistency in supposing that those diseases, in different cases produce both effects: for there is no doubt that tubercles in an incipient state sometimes exist for a considerable length of time, without being accompanied with any morbid symptom. In such cases, any circumstance which can excite inflammation of the lungs must produce such a change in the state of the tubercles as shall render them exciting causes of phthisis. In other cases, the effect of these diseases may be simply inflammation of the lungs, which terminates in suppuration.

THE cause which induces the second species of phthisis pulmonalis seems to be a certain disposition in the blood-vessels of the lungs, to form or to deposite calcareous matter; but the nature of this disposition is too much involved in obscurity to be investigated here.

On the whole, there appears to be three classes of exciting causes of phthisis pulmonalis, corre-

T 2   sponding

fponding to the three fpecies of the difeafe. The
firft comprehends all thofe circumftances which
tend to the production, or perhaps, to fpeak more
accurately, to the inflammation of tubercles; the
fecond confifts of a difpofition in the blood-veffels
of the lungs to form or depofite calcareous matter;
and the third includes every circumftance which
prevents the healing of ulcers formed accidentally
in the lungs.

PROXIMATE CAUSE. It has been generally imagined that the proximate caufe of phthifis pulmonalis is ulceration of the lungs; and this fuppofition has been thought to be founded upon the appearances on diffection *. It has however been already ftated, that, in fome cafes of the difeafe,
there are no ulcerations in the lungs; as, for example, where there is a depofition of calcareous
earth.

Another opinion has been advanced on this fubject, by Dr. Beddoes; viz. that the ftructure of the
lungs is fo altered as to tranfmit a more than ordinary portion of oxygen to the blood †. In fupport of this hypothefis, Dr. Beddoes has endeavoured to fhow (as has been already mentioned),
that

---

* Vide, Home's Princip. p. 147—Cullen's Firft Lines, par. 862.
† Beddoes's Obfervations on the Nature and Cure of Calculus,
Sea Scurvy, Confumption, &c. p. 135.

that the blood in phthifical patients is hyper-oxygenated; and has adduced feveral cafes, to prove that the means moft conducive to the cure of confumption are thofe by which the fupply of oxygen to the fyftem is much diminifhed. His opinion originated from an obfervation, that nature in particular cafes fufpends the progrefs of phthifis pulmonalis, by the very means, which are calculated, he thinks, to diminifh the fupply of oxygen, viz. in cafes of pregnancy. As the fœtus, he remarks, receives from the mother that proportion of oxygen which is neceffary to its exiftence, while at the fame time there is no apparent provifion for an extraordinary fupply of it to the mother; it follows, that, during pregnancy, the fyftem of the woman muft receive a lefs quantity of oxygen than ufual.

Although this ingenious hypothefis be exceedingly plaufible, it is by no means fatisfactory. That pregnancy does not always fufpend phthifis pulmonalis muft be allowed by every practitioner; and that it generally accelerates inftead of retarding the progrefs of the difeafe, has been already alledged by the author of thefe remarks. Befides, the fuppofition of a larger quantity of oxygen than ufual being tranfmitted into the blood, by lungs fo much difeafed as to be incapable in general of admitting above one fourth of the ordinary quantity of air, is inconfiftent with the common opinion
refpecting

respecting the functions of the lungs, the opinion which Dr. Beddoes himself adopts. For if the lungs, in a healthy state, be designed to combine with the blood the oxygen contained in the atmospheric air received into those organs, it is surely unphilosophical to imagine that a diseased state, which prevents not only the admission of the usual quantity of atmospheric air into the lungs, but also the free transmission of the blood through them, should render them capable of combining an increased quantity of oxygen.

AN hypothesis of a different nature, founded too upon chemical principles has been suggested by Dr. Reid, viz. That the diseased state of the lungs prevents the expulsion, during respiration, of the phlogiston and lymph generally thrown off by that operation *. Changing the terms employed by Dr. Reid for those now used, his hypothesis seems to be, That the diseased state of the lungs prevents the admission of oxygen; which is directly opposite to Dr. Beddoes' opinion. An objection, however, immediately occurs against this opinion, which is, That, were it just, phthisical persons should be always cold, in the same manner as the blue boy, mentioned by Dr. Sandyford, the right ventricle of whose heart opened into the aorta,

* Vide, An Essay on the Nature and Cure of Phthisis Pulmonalis, by T. Reid, M. D. first edit. p. 58. et seq.

ta, fo that a fmall quantity of blood only paffed through the lungs; whereas, fuch patients are, on the contrary, always affected with much heat.

What then is to be regarded as the proximate caufe of phthifis pulmonalis? Since the deranged ftructure of the lungs produced by the exciting caufes of the difeafe neither prevents the admiffion of oxygen, nor promotes an increafed fupply of it, at leaft in fo far as has been hitherto proved; may it not be probable that it acts by fubtracting the nutritious part of the blood?

REMARKS on the Cases of Phthisis Pulmonalis.

Cafes of Phthifis a Tuberculis. In the firft cafe, the refemblance between the tubercles and glandular bodies is clearly marked by Morgagni.

The third cafe affords an example of the complication of phthifis and tabes mefenterica. The fimilarity noticed between the tubercles in the lungs, and the protuberating bodies on the furface of the peritoneum, omentum, inteftines, &c. and the fimilitude of thofe bodies to glands, are ftrong circumftances in favour of the opinion that tubercles are of a glandular nature.

The fourth cafe is remarkable from the offifica-

I   tion

tion which had been coughed up from the lungs. It is unfortunate that the hurried manner in which the body was examined prevented the feat of the bone from being difcovered, and alfo rendered it uncertain whether there were in the habit a dif-pofition to form offific matter.

In the fifth cafe, the death of the patient feems to have been occafioned by the fudden rupture of the vena azygos, which was quite an accidental circumftance.

Cafes of Phthifis a Materia Calculofa. The firft cafe was fo well marked, that, before the death of the patient took place, Malpighius fore-told the appearances in his lungs.

In the fecond cafe, there were evident fcrophu-lous tumours throughout the whole body; yet no tubercles appeared on the lungs, but on the con-trary hard ftony concretions. This fhews that, although tubercles do not exift without fcrophula, they are not the neceffary confequences of that morbid ftate. The offifications that appeared within the heart, in the third cafe, fhew a coinci-dence between the difpofition in the veffels to form bone, and to depofite calculous concretions.

§ 2.

————————

§ 2. TABES MESENTERICA.

THIS difeafe begins with irregularity in the ftate of the bowels, fometimes obftinate coftivenefs taking place, and at other times, loofe, flimy, unnatural ftools being difcharged. The appetite for food is very irregular; for fometimes there is a loathing at all kinds of food, and at other times a voracious hunger. The belly is generally fwelled and hard; and where the difeafe occurs in children, it is always fo. In fuch cafes, too, hard knotty bodies often can be felt through the teguments of the abdomen. The whole body is by degrees emaciated. The fkin is generally dry and hot. The mouth is parched, and conftant thirft attends. The uneafinefs is generally aggravated after taking food. Hectic fever, at firft very irregular, and by degrees diftinctly marked, at laft fupervenes. A remarkable appearance in the urine has been noticed, viz. that it is thick, and fometimes chylous *.

Tabes mefenterica moft frequently attacks children; but as the fucceeding cafe fhows, it alfo oc-

Vol. I.   U   curs

* Vide Juncker's Confpectus, pag. 361; from whofe defcription of this difeafe the above hiftory is principally borrowed.

curs in adults. It is fometimes complicated with phthifis pulmonalis.

The appearances on diffection prove that the feat of the difeafe is in the glands of the mefentery. Thofe glands are found enlarged: they are fometimes fo much indurated, as to refemble fcirrhi; fometimes, on the contrary, they are quite foft and flabby; and fometimes they contain pus, mixed with a white foft curdly matter *.

CASE of Tabes Mesenterica.

Case. (xxvii. 16.)

A beggar, aged fifty years, who had formerly been a wool comber, was found dead. He had been hectic for fome time, and had been fo much diftreffed with heat, that although it was the coldeft feafon of the year, he had been accuftomed to lie naked upon fome ftraw in a hut, and in this fituation he was found.

Appearances on Diffection.

Abdomen. A larger than ordinary proportion of fluid was obferved in the cavity of the belly. The ftomach was very much diftended, and on the fuperior part was covered by the omentum, which

was

* Vide Baillie's Morbid Anatomy, pag. 134.

was in a great meafure drawn upwards. The ftomach contained a confiderable quantity of air, and a fmall proportion of urine, with the colour of which the internal furface was tinged. The mefenteric glands,.both in the center of the mefentery where they were collected in the form of a double clufter of grapes of a moderate fize, and alfo in other places here and there, where they were perfectly diftinct, were much larger than u-fual, and fomewhat indurated. Although the liver was of a moderate fize, the fpleen was very fmall; but the fplenic artery was of a larger fize than in proportion to that of the fpleen. The other abdominal vifcera were in a natural ftate.

THORAX. The heart was not fmall, but was flabby. A pretty large bony fcale appeared externally nearly in the middle of its pofterior furface, and a fmaller fcale of the fame kind was feen likewife externally on the right auricle. The internal furface of the aorta, behind the femilunar valves, was marked with whitifh fpots.

CAUSES OF TABES MESENTERICA.

PREDISPONENT CAUSE. It has been generally imagined, that a fcrophulous habit of body is the neceffary predifponent caufe of tabes menfenterica ; and it muft be confeffed that the appearance of the ordinary fubjects of the difeafe, feems to confirm this opinion. Both Sauvages and Dr. Cullen, however, believe that the difeafe fometimes occurs in thofe who have never had any fymptom whatever of fcrophula *.

EXCITING CAUSES. Bad diet and inattention to cleanlinefs, have been commonly regarded as the exciting caufes of this difeafe. But it is not eafy to underftand the exact manner in which thofe circumftances can produce fwelling, induration, or fuppuration of the mefenteric glands.

PROXIMATE CAUSE. The obftruction of the mefenteric glands, by which a due fupply of chyle is prevented, is obvioufly the proximate caufe of tabes mefenterica.

ANOMALOUS CASES OF TABES.

CASE I. (XXIX. 12.)

A WOMAN, aged forty years, who had been accuftomed to live chiefly upon falt victuals, and to drink plentifully of wine, had been for many years

* Vid. Sauvages Nofol. Med. tom. 2. p. 449; and Dr. Cullen's Firft Lines, par. 1556.

years fubject to pains of the ftomach. Impaired
appetite for food, naufea, and foon after, repeated vomiting of blood, together with conftant fever, reftlefsnefs, and thirft, fupervened to thefe
pains. Although no hardnefs could be perceived
in the belly, yet a certain uneafy fenfation was
from time to time felt in the region of the ftomach during the abfence of the fevere pain, even
though no preffure were made. She complained
alfo of pain in her loins; but it was only when fhe
had been working more than ufual, or had been
carrying fome burden. Along with all thofe
complaints, fhe was fometimes affected with a very fevere pain in the head. Whenever thefe fymptoms became violent, more efpecially the ftomach
complaints, fhe received confiderable relief from
blood-letting. She feemed, too, to derive much
benefit from drinks, in which a piece of bread only had been boiled. She appeared alfo, more
than once, to have become convalefcent from the
daily ufe of milk diet, and her fpirits were kept
up by the regular appearance of the menfes,
which flowed at ftated periods, till the time of
her death.

At laft a hard tumor appeared on each fide a-
bove the clavicles, at that part where the external
jugular vein paffes. Thefe tumors occafioned
confiderable pain, and as they did not yield to a-
ny remedy, but on the contrary, increafed daily,
they

they rendered the refpiration difficult. Along
with thofe fymptoms, fhe had conftant fever,
which, preceded by a flight chilly fit, increafed
always towards the evening. She complained of
pain in her head; and the pain in her ftomach
was conftant, but was not now attended with vomiting of blood. She was perpetually diftreffed
with thirft, and had a fenfation of much bitter-
nefs in the mouth; from which organ, during the
latter days of her life, a very fetid fmell proceeded; but no pus was ever obferved to be fpit up.
Under thefe circumftances, fhe dragged out a
miferable exiftence, much longer than from the
fmall and feeble ftate of her pulfe could have been
expected. Her pulfe for the laft fifteen days, during which time fhe took nothing but a little broth
and fome wine, became fmaller and more frequent. At length fhe died.

Appearances on Diffection.

EXTERNALLY. The carcafe was very lean.

ABDOMEN. The omentum was rolled up towards the upper part of the abdominal cavity, in
fuch a manner, that the tranfverfe arch of the
colon, which commonly lies under the ftomach,
appeared immediately under the umbilicus. It
might have been forced into this fituation partly
by the ftomach, as the left part of the fundus of
that vifcus defcended lower than ufual. The ftomach externally, and particularly at one place to

a

a large extent, was of a livid colour. Its coats were thickened and indurated; except in certain parts, where they were fo putrid that they were lacerated on touching them. Through thefe lacerations a very fetid cineritious matter, like a kind of poultice, which was contained within the ftomach, was difcharged. This fluid had burft into the ftomach from a tumor or abfcefs of the worft kind, which had been fituated on its pofterior furface: for at that place the ftomach was to a great extent immoderately thick, fwelling inwards; unequal in its furface; in a filthy, rotten, and gangrenous ftate; and of the fame colour as the above mentioned matter. The pylorus was found. All the inteftines, the colon not excepted, were in a contracted ftate; which was to be expected after fo long continued lofs of appetite. The fpleen, although found, was of a larger than ordinary fize, and was internally of a pale colour. On the right fide of the liver fome fcirrhofities, of a white colour, and roundifh form, like common fized grapes, were obferved. Thefe were difpofed over the furface in fuch a manner, that they were in fome degree concealed within the fubftance of that organ. In cutting through the liver, one of thefe tumors, completely buried within its furface, appeared. The gall-bladder contained a large quantity of very yellow bile, which had tinged the neighbouring parts. On the pofterior furface of the left kid-

3   ney

ney there was an oblique line, of a confiderable
length, of a whitifh colour, and of an apparently
tendinous fubftance. This line penetrated to fuch
a depth that it reached the tubulæ in which the
papillæ are received. It had very much the appearance of the cicatrix of a former ulcer; but no
mark of this could be difcovered, neither in the
furrounding adipofe membrane, nor in the abdominal mufcles. The uterus was fmall, and fituated very low; and inclined fo much towards the
right fide, as to be greatly nearer that than the
left fide. The round ligament, too, was much
fhorter on the right fide than on the left. The
cervix uteri, and more efpecially the os tincæ,
were found nearly in the fame ftate as in virgins;
the former having, internally, the oblique and
prominent rugæ, which characterife it; and the
latter having a narrow and round aperture. The
hymen, though fmall, was diftinct, and exhibited
no marks of laceration; but within the hymen,
neither were the carunculæ myrtiformes obferved,
nor were the rugæ of the vagina diftinctly marked. The fkin too, at the lower part of the abdomen, being indented with whitifh coloured pits,
did not correfpond with the appearances of the
cervix uteri and hymen. The ovaria were of a
large fize, confidering the age of the woman and
fmallnefs of the uterus. Externally they were
wrinkled: internally, the left contained fome fmall

empty

empty cells included within a thick white membrane; and the right had, within a cell not much larger than thefe, fome black coloured half coagulated blood. The right Fallopian tube was pervious to the ovarium, but impervious towards the uterus: the left, on the contrary, was open only towards the uterus. A confiderable quantity of fat was obferved in the mefentery, and in the interftices of the mufcles of the back and limbs; and a fmall quantity alfo in the omentum. This was furprifing in fo lean a fubject; but indeed it was a female one. The abdominal mufcles were of a very beautiful red colour. Some glands, greatly enlarged, lay hid beneath the yellow fat contained in the mefentery, which covered the lumbar vertebræ and the trunks of the large veffels contiguous to them. Thefe glands adhered fo clofely to thofe veffels, as to be feparated with great difficulty. They were internally white, and not very hard; and they contained purulent ichor. The other glands throughout the mefentery were not enlarged. Near the ftomach, however, one of the lymphatic glands was obferved to be much thicker and harder than ordinary: it was of a dirty yellow colour. The whole of the pancreas had become thickened, and at the fame time fomewhat dry and hard; except in one part, which grew out into a white fubftance refembling that of the thymus gland.

Vol. I.   X   Thorax.

THORAX. The two loweſt jugular glands wcre of a white colour, and were enlarged to the ſize of two inches at leaſt. Theſe glands conſtituted the hard tumor on each ſide under the clavicles, mentioned in the hiſtory of the caſe. They were found to be conſiderably hard, although they contained a purulent ichor: part of this flowed out when the clavicles, under which, and on the contiguous part of the ſternum, they lay, werc removed. The other jugular glands alſo were in the ſame ſtate, with reſpect to colour, hardneſs, and contents; but had not increaſed to ſo large a ſize. The axillary glands had undergone no change whatever. But thoſe glands placed at the firſt diviſion of the aſpera arteria, were of a white colour; and, from a ſmall ſize, had become not leſs than ordinary grapes. They were ſomewhat hardened too; and contained the ſame kind of purulent ichor ſeen in ſo many other glands. The aſpera arteria, however, at leaſt towards thc throat, was found; as was alſo the œſophagus in its whole length. The lungs, which were turgid with air, werc free from diſeaſe. A number of roundiſh tubercles, of a hard compact ſubſtance, and of a depreſſed figure, ſo numerous as to be almoſt contiguous to each other, beſet all the edges of the valvulæ mitrales. In one of the ſemilunar valves a ſmall ſcale, not however bony, was obſerved. In other reſpects the hcart was found.

HEAD.

HEAD. The brain not only was not flabby, but even approached towards hardnefs : it was in a natural ftate, except that a fmall quantity of pellucid fluid was found in the lateral ventricles, and that the plexus choroides were pale. The pineal gland was fomewhat more firm and globular, and of a whiter colour, than it ufually is.

CASE II. (XLVII. 4.)

AN unmarried woman, who had for many years had no menftrual difcharge, and who had been long troubled with ulcers in the legs, having become hectic, died.

Appearances on Diffection.

ABDOMEN. A confiderable quantity of effufed fluid was found in the belly. On the internal furface of the uterus, a great number of protuberant apparently glandular bodies were obferved ; there were few, however, towards the fundus. The ovaria contained no veficles, and confifted of a whitifh fubftance fimilar to that of the pancreas, except that it was of a fofter confiftence.

THORAX. A quantity of effufed fluid was found in the cheft, as well as in the belly.

CASE III. (XLIX. 16.)

A MAN who was greatly emaciated, and was by

X 2   fome

fome deemed affected with phthifis, was brought into the hofpital of Padua, where he died.

Appearances on Diffection.

ABDOMEN.   All the vifcera were found.

THORAX.   The lungs and other contents of the cheft were in a natural ftate.

HEAD.   The dura mater was very much thickened, and the brain was very flabby. It was found, on examining the fuperior part of the medulla fpinalis, that the dura mater could not be fo eafily feparated as ufual; and that, in drawing it off from the contiguous membrane, much caution was neceffary in order to prevent laceration.

The fkin in this fubject was very hard, as it commonly is in tabid bodies.

C A S E   IV.   (XLIX. 18.)

AN old man, who was very much emaciated, died, it was faid, in confequence of the marafmus incident to old age.

Appearances on Diffection.

ABDOMEN.   The mefenteric glands were not fo fmall as they generally are at that age. A great number of glands, of a large fize, both with refpect to thicknefs and to length, fome of them being two or three inches long, were obferved about the iliac veffels from their origin quite down to the thighs. Thefe were placed in fuch a manner, that

they

they feemed to cover the iliacs, like a continued chain ; and prefled upon them fo much, that the parietes of thofe veffels appeared fomewhat in-flected and varicofe.   When cut into, thefe glands did not appear! different from the natural ftate of lymphatic glands.   The fpleen, though rather fmall in its other dimenfions, was thicker than u-fual, efpecially about the middle.   The membra-nous bands by which it is connected to the dia-phragm were alfo thicker than common ; and be-fides, its coat was not only likewife thick, but was alfo, in the centre of its convex furface, indurated for a fpace equal to a circle of two inches diame-ter, and in fome part of that fpace it had become offified.   Internally, a trunk of the veffel, alfo in-durated, belonging to the fubftance of the fpleen, appeared connected to the fame part.   The fple-nic artery, for the extent of fome fingers breadth from its origin at the cœliac, was fomewhat nar-rower than natural ; but at the part where it begun to be contorted, as ufual, it became wider. The urinary bladder, which was fo much diftend-ed with urine that it appeared above the pubis, had its coats thickened.   When comprefled by the hand, the contained urine was not eafily dif-charged ; nor was it poffible by thefe means to force out the whole of it.   This was probably oc-cafioned by the ftate of the proftate gland ; for that body was enlarged, and protruded from the

internal

internal orifice of the urethra into the cavity of the bladder. It was externally of a brownifh red colour. Within its fubftance, which was in other refpects in a natural ftate, granules of a blackifh yellow colour, like tobacco, were feen in feveral places.

THORAX. The heart was completely deftitute of fat; its furface was of a dirty yellow colour, and was not fmooth. The valves of the aorta were very hard: within the internal furface of that, veffel itfelf, beyond the valves, large bony fcales were obferved. But in that part of the aorta fituated in the belly, and in the iliacs continued from it, incipient offifications, or white fpots, only were found.

The carotid arteries had a fingular appearance; having afcended one half of their height, they became contorted like a fcrew, and immediately after returned to their original ftraitnefs; and fo obftinate was that contortion, that, although the arteries were drawn out into a ftraight line, they affumed their original form the inftant they were left to themfelves.

CASE V. (LXV. 3.)

A MAN, aged forty-four years, having been at a diftance from his own home, in a mountainous and uncultivated country, had taken, on account

of

of a flight gonorrhœa, many mercurial medicines. Although thefe medicines were probably ill prepared, he could get no other in that country. While ufing thofe remedies, his ftomach was very often irritated, and vomiting frequently took place. From that period, he began to vomit almoft every thing which he eat or drank; and if he did not vomit, he became very much diftreffed with a pain in his ftomach, which he felt at all times in a flight degree; and alfo with hiccup. When he took food immediately after vomiting, he moft generally retained it. He had a large difcharge of thick ill-fmelling faliva. His belly was coftive; and when glyfters of milk were ufed, nothing was expelled but hard excrementitious lumps. Although the pulfe was not at firft affected, yet there was a confiderable emaciation of the body from the beginning. Many different medicines were employed, but in vain; for he at laft died.

Appearances on Diffection.

ABDOMEN. The pylorus of the ftomach was very narrow and very hard. Near it there was a fmall ulcer; on the other parts of the internal furface of the ftomach feveral bodies like glands were fcattered here and there.

C A S E VI. (LXV. 5.)

AN old woman, who had been long affected with

with an obftinate diarrhœa, attended with a great
lofs of flefh, became at laft reduced to a ftate of extreme weaknefs; and in this fituation died.

Appearances on Diffection.

ABDOMEN.   Inftead of the adifpofe membrane
placed under the fkin, which is very rarely wanting even in the moft emaciated women, a thin
membrane like net-work, fcarcely retaining the
fmalleft veftige of fat in any part of it, was obferved.   In the belly neither was there any ill fmell,
nor could any remarkable appearance at firft fight
be difcovered, except that the bile in confiderable quantity and of a deeper than ordinary colour, contained in the gall bladder (which was
pretty large, and extended confiderably below the
liver) had tinged its contiguous parts more extenfively than common.   Neither the fmall inteftines, nor the ftomach, which was in a contracted
ftate, were in any degree difeafed.   Some parts
of the internal furface of the large inteftines, efpecially about the cœcum and colon near the val-
vula coli, were red from inflammation.   The rectum was quite livid from previous inflammation,
and on its internal furface it was in feveral places,
efpecially towards the inferior part, fwelled out.
On the fwelling at the lower part, a fpot was obferved extending upwards the length of a finger's
breadth.   This fpot was foft and prominent, appearing as if it were formed from fome half coa-
                                                    gulated

gulated blood placed under the internal coat of the inteftine. Above this part, feveral bodies, either true lenticular glands, or of a fimilar nature, of a reddifh or rather brown colour, appeared difperfed up and down. Some of the mefenteric glands, although of a found texture, feemed of a larger fize than is common at that age. The trunk of the aorta was fomewhat hard in feveral places, and on its internal furface was of a whittifh colour, in confequence of many incipient offifications. On the internal furface, too, the pofterior and lateral portions of the veffel formed, by their junction, an angle inftead of a curved line.

CASE VII. (LXX. 5.)

A PORTER, aged fifty years, who had been greatly addicted to drinking, and accuſtomed to indulge in much eating, was affected, three months previous to his death, with a tertian fever. This man, not only had not been a valetudinarian, but had even been in perfect health, except, as was learned from his companions, that now and then he complained of a pain fituated in the epigaftric region, about the fcorbiculus cordis. When the fever began to abate, he became fubject to vomitings; which difappearing for fome time again, returned and continued to trouble him during the reft of his life. He never vomited any thing but

VOL. I.   Y   his

his food; and when vomited it had never any bad
fmell or tafte. During the laft month of his life,
he could retain nothing on his ftomach but muf-
cadine wine, which was given by way of cordial.
In confequence of this, he became hectic, and had
a great wafting of flefh. Every medicine, both internal and external, that was employed to check
thefe vomitings, proved ineffectual. Among thofe,
quickfilver was ufed, to the quantity of three ounces, about a month before his death, when an obftinate coftivenefs afforded reafon to apprehend i-
liac paffion. This medicine neither proved of a-
ny advantage, nor was afterwards feen during the
operation of clyfters; but it was fuppofed by the
attendants, to have been afterwards difcharged a-
long with the fæces, when the ftricture of the
bowels was removed. Under the fymptoms above
ftated, he died, never having complained of any
tumor or pain in the belly; his pulfe alfo having
never been much affected, except only that it
now and then intermitted.

Appearances on Diffection.

EXTERNAL APPEARANCES. The body was fo
exceedingly emaciated, that on the back of the
hands, and on the upper part of the feet, the
bones of the metacarpus and thofe of the meta-
tarfis could have been very well demonftrated;
and the eyes were, in confequence of the deficiency of fat in the pofterior part of the focket, moft
                                  aftonifhingly

aftonifhingly funk. From the lobe of the left ear, a flender brafs ring depended, which is common- ly the mark of a previous diforder in the neigh- bouring eye, but both eyes were in a found ftate. As the mouth happened to be open, it was ob- ferved that there were but few teeth in the jaws.

ABDOMEN. The inteftines were here and there marked with livid fpots. They were unufually contracted, the natural confequence of the daily vomiting. The large inteftines, however, were contracted for a fhort fpace only, as they contain- ed fæces, though not in a large quantity. The ftomach was in an unufual fituation, of an uncom- mon length, and had a very peculiar pofition : For, beginning at the ufual place, (no part of the œfophagus having paffed into the belly) it pro- ceeded through the left fide of the belly, in a ftrait line, down as far as the pubis; from whence it turned upwards towards the right fide, and ter- minated in the duodenum. It was of a moderate breadth; and contained nothing but a thin poul- tice-like mafs, which was fuppofed to be the re- mains of the little food that had been taken in. When the pylorus was handled, externally a con- fiderable hardnefs was felt. On opening the fto- mach, the ring of the pylorus was obferved to be divided as it were into two or three protuberances, which, although not large, were hard. That part of the ftomach next the pylorus was indurated to

the

the extent of two fingers breadth; the coats being there thickened, and in a ſtate approaching
to the hardneſs of bone. They were not really
oſſified, nor did they ſtraiten the paſſage; but,
in conſequence of their unyielding ſtate, they had
been incapable of propelling the food into the
duodenum. For a ſhort ſpace from the indurated
portion, the internal ſurface of the ſtomach was
ſlightly livid. It is probable that the ſtomach
contained quickſilver, which had flowed out when
that organ had been moved to one ſide; for it
was found, to the quantity of an ounce and a
half, within the duodenum, which had been tied
up a litlle below the pylorus. It was not obſerved in any of the other inteſtines. The urinary
bladder contained little urine, and was in a ſound
ſtate; as was alſo the urethra; but no veſtige of
the ſmall, oblong, oval protuberance, generally
placed at the beginning of the urethra, could be
traced at that part. The ſlender line which terminated as uſual in the verumontanum appeared.

THORAX. The heart, as might be expected,
was deſtitute of fat; but, what was very remarkable, more eſpecially in a man who was of a tall
ſtature, who had been always employed in a laborious manner of living, and who had been accuſtomed to lift heavy burdens, it was ſo ſmall,
that it appeared rather to be the heart of a child
than that of a man. All its dimenſions, as well

as

as the thicknefs of its parietes, were in exact proportion to the fmallnefs of its fize. No difeafe
appeared, externally nor internally; except that
the fuperficial veins were in fome parts varicofe.
Although the trunk of the aorta feemed to be
dilated, the valves were in a natural ftate. The
diaphragm had become offified at the right fide
of the centrum tendineum; or at leaft, at that
part, a bony lamina, of fome thicknefs, meafur-
ing an inch and an half in length, narrow at one
of the extremities, and becoming gradually wider
at the other, but no where exceeding a finger's
breadth in width, was placed between the pleura
and peritoneum.

HEAD. A fmall quantity of fluid was found in
the lateral ventricles. On the choroid plexuffes,
within the lateral ventricles, where they are reflected upwards at an angle to cover the thalami
nervorum opticorum, a pretty large hydatid, appearing at firft like mucus, was obferved. The
pineal gland feemed very fmall: it contained
a corpufcle, which, in proportion to the fize of
the gland, was not inconfiderable. This little body was of an irregular figure; was hard, not friable, and was of a dirty yellow colour. The medulla fpinalis was very flabby.

CASE

C a s e VIII. (xxxix. 14.)

A SLENDER woman, of about forty years of age, who had been much fubject to hyfteria, and particularly to violent hyfteric paroxyfms, which affected her whole body, and more efpecially the vifcera of the abdomen, with convulfive motions, after having had fome fits that were more violent than ufual, began to obferve an evident depreffion in the epigaftric region, and at the fame time a fwelling in the hypogaftric. The depreffion never changed; but the fwelling often altered within the fpace of a fingle day; for, although it appeared large and very hard, it often fuddenly fubfided. When fhe took food, fhe obferved that it defcended into that part and increafed the fwelling, and alfo rendered the fenfation of weight which fhe always felt there more uneafy. At the diftance of four or five hours from that time, fhe ufually became affected with violent pains, tormina, and faintings. She often complained that all her bowels had fallen out of their proper fituation, as fhe expreffed it. Her digeftive powers were obvioufly impaired. She was feverifh, and was much emaciated. Having lived in this manner for three months, fhe died.

Appearances on Diffection.

ABDOMEN. The ftomach was found to have
fallen

fallen down into the epigaſtric region; ſo that there was ſcarcely four fingers breadth between it and the pubis. That part which is connected to the œſophagus was ſo much elongated, that the whole fundus of the ſtomach lay in the epi-gaſtrium.

C A S E IX. (xl. 23.)

An old man, apparently ſixty years of age, who was ſo very beggarly that he was forced to pick up a miſerable ſubſiſtence from the outer rinds of melons, or any other ſubſtance thrown into the ſtreet, having become affected with fever, toge-ther with a ſenſe of oppreſſion in the cheſt, at-tended by difficult reſpiration, weak pulſe, con-ſtant cough, and the expectoration of catarrhous matter, was received into the hoſpital of Bologna, where he had formerly repeatedly been. After he had felt himſelf conſiderably relieved, he went out again into the ſtreets; but within a ſhort time he returned into the hoſpital. He was then ſo much emaciated, and ſo much exhauſted by diſ-eaſe, cold, and hunger, that he died ſoon after his admiſſion.

Appearances on Diſſection.

Abdomen. The ſtomach was large, although it contained almoſt nothing. Internally it had no rugæ; and was of a browniſh colour, here and
there,

there, for a confiderable fpace, from the middle
towards the left fide, and efpecially towards the
œfophagus, where the brown colour penetrated
more deeply. The whole convex furface of the
liver, except a fmall fpace at the lower part on the
right fide, adhered firmly to the feptum tranfverfum. In that part, the furface was hollowed out
by an hydatid of the diameter of a finger's breadth.
The membrane of the convex furface of the fpleen
was in one part of a white colour, and the middle
of that part was become bony to a fmall extent.
The fpleen itfelf was of a fofter confiftence than
ufual, and was larger rather in thicknefs than in
length or breadth. The fplenic artery, however,
appeared wider than even that increafed thicknefs required. The mefenteric glands were very
diftinct, although in a man of that age. Many of
them were of the fize of a bean; but, when examined accurately, they were found to be certainly
free from difeafe. The fize of the kidneys was
unufually fmall in proportion to that of the body.
The appearance of their furface was uncommon;
for they were equally convex on the pofterior as
on the anterior furface. Both furfaces were unequal, and in fome degree knotty; and more efpecially in the left kidney. That kidney too exhibited certain depreffions as if from cicatrices. The
orifices of the ureters within the bladder feemed
larger than ufual. The bladder itfelf internally

was of a red colour; and was here and there marked
with fanguiferous veffels, which were as diftinct as if
they had been filled with coloured wax. Externally it was furnifhed with redder fibres than ufual.
The iliac arteries were tortuous, as the fplenic artery generally is. The iliac veins, as far as their
divifion, were fo much corrugated, as it were, that
they could be extended with difficulty.

THORAX. Within the thorax, as well as in the
pericardium, a fmall quantity of watery fluid was
found. The lungs were attached firmly to the
fides and to the back. The right lobe, when
drawn away, left a kind of opaque, thick, firm,
uniform coat, adhering to the parietes of the thorax, extending from the lower part to beyond the
middle of its length, and from the fpine almoft to
the fternum. This membrane, when pulled by
one extremity, and by the part attached to the
fternum, was drawn off entire. It neither belonged to the pleura, nor was it the membrane of the
lungs; for both thefe parts remained in their natural fituation. It was therefore probably a membrane formed in confequence of inflammation.
The lungs themfelves were not very found. In the
upper part of one of the lobes a hard fubftance was
perceived. The heart was twice as large as it naturally is: it contained no blood, but only a few
moderately fized polypous concretions. Both ventricles were dilated. The right ventricle, and the

VOL. I.   Z   corref ponding

correfponding auricle, which was alfo much dilated, had very thin parietes ; on the contrary, the
parietes of the left ventricle were thicker and
harder than ufual. The valvulæ mitrales were
enlarged, and their lower edges were much thickened and knotty. The figmoid valves were not
fo foft as ufual; and the femilunar were ftill
lefs' yielding; one of them being already bony at
one part of its lower circumference. The aorta,
before its curvature, was wider than ufual. The
whole of its internal furface was marked here and
there with white fpots; and the fame appearance was obferved within the iliacs. Some of
thefe fpots appeared prominent on the internal
furface, and were very hard and bony. This was
more efpecially the cafe at that part from whence
one of the inferior intercoftals went off. The orifice of that veffel, happening to lie in the center of
the fpot which protuberated in a circular form,
had been in confequence fo ftraitened, that together with the fpot, it at firft fight appeared like a
large lenticular gland.

CAUSES of Hectic Fever.

Predisponent Cause. Daily obfervation proves
that perfons of a fcrophulous habit are chiefly predifpofed to hectic fever ; but, as it is well known
that

that thofe of a different habit are alfo liable to the
difeafe, fcrophula cannot be confidered as the fole
predifponent caufe.   In every cafe, the difeafe is
preceded by general debility of the fyftem. It may
therefore be probable, that the effects of the exciting caufes cannot take place unlefs the body be
in that ftate.   It muft be confeffed, however, that
this mode of reafoning is not perfectly fatisfactory.

EXCITING CAUSES.   An author of the higheft
eminence has alleged, that the abforption of purulent matter is the fole exciting caufe of hectic
fever* ; while another, of confiderable refpectabi-
lity, has denied that the difeafe is ever produced
(at leaft in phthifis pulmonalis) by that caufe†.
Both are certainly miftaken : for, unlefs the evidence of the fenfes be entirely laid afide, it cannot
be doubted that hectic fever is very often induced
by the abforption of pus; but, on the other hand,
it muft be alfo allowed that the difeafe fometimes
occurs where no pus could be abforbed.

Suppuration in the lungs, in the liver, in the mefenteric glands and other abdominal vifcera, in the
external parts of the body, and throughout the
whole fyftem, as in cafes of fmall pox, &c. often act
as exciting caufes of this difeafe ; but, as they do not

Z 2   .uniformly

---

* Dr. Cullen's Firft Lines, par. 861.

† Vide, An Effay on the Nature and Cure of Phthifis Pulmonalis, by T. Reid, M. D. firft edition, page 59, et feq.

uniformly produce the effect, it has been suppofed that fome peculiar ftate or condition of the matter abforbed is requifite for this purpofe *.

In thofe cafes where this fever cannot be traced to the abforption of purulent matter, all the various circumftances from whence the difeafe can be thought to originate, tend uniformly to produce one effect, viz. the prevention of a regular fupply of chyle. Thus, obftructions of the mefenteric glands, indurations of the ftomach and other chylopoetic vifcera, exceffive habitual evacuations, great irregularities in diet, as frequent drunkennefs, and long continued affections of the mind that impair the appetite for food, as immoderate grief, which have been long acknowledged by practitioners to be occafional exciting caufes of hectic fever, agree only in one refpect, that of preventing the ordinary fupply of chyle.

On the whole, therefore, it is prefumed that the abforption of purulent matter, and every circumftance which tends to prevent the formation or the fupply of chyle, occafion this difeafe.

Proximate Cause. From the phenomena of hectic fever, the proximate caufe has been generally fuppofed to be irregularity of action of the vafcular fyftem in confequence of an acrimonious
ftate

---

* Vide, Dr. Cullen's Firft Lines, par. 861.

state of the blood. This opinion, however, is neither satisfactory to the theorist, nor useful to the practitioner. As the qualities of the blood are not yet accurately known, it may indeed be urged that it is not easy to employ any other epithet than that of acrimony, for expressing the particular condition of that fluid which excites irregular action of its containing vessels. It is probable, however, that in the present instance a more accurate expression may be adopted. That the blood cannot perform its natural functions, unless it be constantly supplied with oxygen and with chyle, is generally believed; and that hectic fever is occasionally produced by circumstances which prevent or diminish the supply of the latter fluid, has been proved: May it not therefore be concluded, that the morbid state of the blood, in the disease under consideration, consists in a deficiency of chyle? If this were admitted, it would follow as a consequence, that the absorption of purulent matter, where it proves the exciting cause, acts by destroying the properties of the chyle. As an objection to this explanation, it may perhaps be alleged, that the disease is commonly aggravated after meals; whereas, were the hypothesis now offered just, it should be always alleviated at those times. In reply; it may be observed, that the circumstance which furnishes this objection has not been universally acknowledged by practition-

ers;

ers *; and therefore cannot be affumed as a principle.

REMARKS on the Anomalous Cases of Tabes.

THE firft, fifth, feventh, eighth, and ninth cafes, afford ftrong prefumptive evidence in favour
of the idea, that hectic fever is immediately produced by a deficiency in the fupply of chyle;
for in all of them the principal morbid appearance
occurred in the ftomach. Thus, in the firft and
fifth cafes, that organ was ulcerated; in the eighth,
it was enlarged, and had its fituation altered; in
the ninth, its internal furface was difeafed; and,
in the fifth and feventh, the pylorus was indurated. In the fixth cafe, it is probable that the difeafe was occafioned by the long continued diarrhœa; which gives additional fupport to the fame
opinion.

Could the fwelling and fuppuration of the lymphatic glands, which were noticed in the firft cafe,
proceed from the ulceration of the ftomach?

The fecond cafe is an example of hectic fever
complicated with dropfy.

The caufes of the difeafe in the third and fourth
cafes are quite obfcure.

CHAP.

* It is denied by Dr. Cullen, Firft Lines, par. 859.

CHAP. II.

INFLAMMATIONS.

General Observations on Inflammation *.

INFLAMMATION † confifts of an increafed action of the arteries, together with fuch an aug-
mentation of blood within them as occafions the
fenfation of heat and pain.   An external part of
the body therefore is faid to be inflamed, when
the action of its veffels is increafed, and when it is
red, hot, painful, and fwelled.   Although inflammation be local, yet if it be confiderable, or feated

* In this part of the work, inflammation fituated within the
head, neck, cheft, and belly, is alone to be confidered. Previous
to that, it appears neceffary to offer fome obfervations on the nature of inflammation in general.

† On the fubject of inflammation the reader may confult the
Works of the late Dr. Whytt, page 211 ; M'Bride's Works, 4to
edit. page 137 ; Dr. Cullen's Firft Lines, par. 2, 35, &c.; and an
Effay by Dr. Carmichael Smith, in Med. Communications, vol. ii.
page 168, from which the principal obfervations here offered are
borrowed.

ed in internal organs, the action of the whole vascular fystem is accelerated: confequently the inflammation of an internal part is indicated, by pain and interrupted function of the affected organ, increafed heat of the whole body, and accelerated action of the vafcular fyftem.

The phenomena of inflammation are different, according to the nature of the exciting caufe, the function or ufe, of the part infiamed, and the natural or acquired texture of the fame.

The inflammation produced by many of the exciting caufes, although left entirely to nature, frequently terminates favourably, running a certain courfe; while that excited by others cannot be overcome by the natural powers of the conftitution.

The function or ufe of the part inflamed influences the phenomena very confiderably. Thus it is obvious that the inflammation of the ftomach and bowels muft be attended with fome fymptoms which do not appear in that of the hands or feet.

The natural texture or ftructure of the inflamed part occafions much variety in the phenomena. The inflammation of the fkin, of the cellular membrane, of mucous membranes, of diaphanous membranes, and of mufcular fibres, is accompanied in each cafe with different fymptoms.

The laft circumftance mentioned as influencing

I   the

the phenomena is, the texture of the organ having been previously altered by the difeafe. The
beft illuftrations of this occur in the cafes of fcrophula and tubercles in the lungs.

Befides thefe circumftances, which are enumerated by Dr. Carmichael Smith, it is probable that
the phenomena of inflammation are affected alfo
by the ftate of the veffels that conftitute the difeafe.

The feat of inflammation is the arterial fyftem
chiefly; but the incipient branches of the veins
alfo generally appear unufually diftended; and
the capillary veffels, and fometimes too the lymphatics, are loaded with blood.

CAUSES of INFLAMMATION.

PREDISPONENT CAUSE. Although it be very
obvious that the application of certain exciting
caufes produces inflammation in almoft every variety of habit; yet it cannot be doubted, not only that fome perfons are more fufceptible than o-
thers of the impreffion of thofe caufes; but, alfo,
that there are particular ftates of the fyftem, which
render an individual more readily affected by
them at one time than at another.

Perfons of a rigid fibre, and of a fanguine temperament, and thofe who indulge much in the

ufe of animal food and ftrong fermented liquors, are particularly liable to inflammation. Plethora feems a great predifponent caufe; but it may be the effect of a peculiar ftate of the veffels, which ought rather to be confidered as fuch. Women are lefs predifpofed to inflammation than men.

Befides the circumftances which render an individual fubject to inflammatory complaints in general, there muft be others which determine the feat of inflammation. Thus, of two perfons expofed to cold, one fhall have cynanche tonfillaris, and the other pneumonia. This muft depend upon fome ftate of the veffels of the affected part; but the precife nature of that ftate is involved in much obfcurity. Some authors have imagined that it confifts in an increafed tone or contractility of the mufcular fibres of the arterial fyftem *. Others, on the contrary, regard it to be quite the reverfe. The moft plaufible argument in favour of the latter opinion is, that parts which have been once inflamed are exceedingly apt to be again fimilarly affected. Still, however, the queftion does not appear to be fatisfactorily folved.

EXCITING CAUSES. Every circumftance which either renders the part acted upon unufually affected by its ordinary ftimuli, or increafes the
number

---

* Vide Dr. Cullen's Firft Lines, par. 247.

number or power of the stimuli applied to it, has been regarded as an exciting cause of inflammation *. This is a very ingenious arrangement, but a more minute one is necessary for the explanation of the phenomena.

All the exciting causes of inflammation may be arranged under the following heads: Mechanical Stimuli, Chemical Stimuli, or those which do not act mechanically, an Increased Degree of the Ordinary Stimuli, Diseases of the General System, and Certain Degrees of Cold.

1st, *Mechanical Stimuli*, include blows, whether attended with division of parts or not; and obstruction to the course of the blood, from pressure, &c. where it excites pain, or irritates the vessels.

2dly, *Chemical Stimuli* comprehend not only chemical stimuli properly so called, as caustics, &c. but also all the stimuli, the operation of which cannot be explained on mechanical principles. Such are certain poisons generated in the bodies of animals, either naturally or morbidly; as the venom of insects, and what is termed the venereal virus, &c.

3dly, *An Increased Degree of the Ordinary Stimuli of a part*. Thus light produces inflammation

A a 2   of

---

* Vide Tentam. Med. Inaug. quædam de Inflammatione complect. auctore Rich. Fowler, Edinburgi 1793, p. 4.

of the eyes, when applied in a degree to which the eyes have not been accuftomed.

4thly, *Difeafes of the General Syftem.* Under this head may be claffed fevers, and cafes where the fluids are impregnated with poifonous matter, fuch as the contagion of fmall-pox, &c. and alfo cafes where the component parts of the blood are not in due proportion to each other.

Laftly, *Certain Degrees of Cold.* It is not eafy to explain the precife degrees of cold which produce inflammation. Experience fhews that thofe degrees vary according to the previous ftate of the fyftem. Thus, if the body be paffed fuddenly from a very warm temperature to a cold one, as for example, from one of feventy degrees of Farenheit, to one of thirty-five or forty, inflammation will be induced. But if the change from thefe degrees be made gradually, no fuch effect follows. Some degrees of cold are productive of mortification; and it has been fuppofed, that the degrees immediately preceding thofe in the fcale excite inflammation.

The heat of the human body may be diminifhed either by the temperature of the air by which it is furrounded, or by the application of moifture of a lower heat than that of the body.

It has been alleged that the fubtraction of heat, or according to common language, the application of cold, caufes inflammation, by ftopping the

the perfpiration; but it is probable that it generally produces the effect of exciting the veffels into unufual action.

PROXIMATE CAUSE. Although many theories
have been propofed on this fubject *, three only
feem to merit any attention. *Firft*, That along
with increafed action of the blood-veffels there is
a fpafmodic ftricture on their extremities; *Secondly*, That, befides the action of the veffels being
increafed, the refiftance to the courfe of the blood
is diminifhed; and *Thirdly*, That the action of the
veffels is impaired, from their coats being in a
debilitated or paralytic ftate.

The firft of thefe theories is adopted by Hoffman and Dr. Cullen. The chief arguments in its
favour are, that every confiderable inflammation
is preceded by a cold fit, and is accompanied with
other fymptoms denoting fpafm on the extreme
veffels; and that the moft effectual means for the
cure of inflammation are thofe which are calculated to moderate the action of the veffels, and to
induce relaxation over the whole fyftem. But as
the beginning veins of an inflamed part are in a
ftate of over-diftenfion, as well as the arteries, it
is

* A view of the various hypothefes of inflammation is exhibited in Meza Compend. Med. Pract. Fafcicul. alt. Inflam. et Dolor.
Sift. in præfat.

is evident that there can be no fpafmodic ftricture, fuch as is fuppofed by this hypothefis.

The fecond opinion has been párticularly recommended to notice by Dr. M'Bride *. That the action of the arteries in the affected part is encreafed, cannot poffibly be doubted : it is proved not only by the nature of the exciting caufes, but alfo by the phenomena of the difeafe. The circumftances which feem to fhow that the refiftance to the courfe of the fluids is diminifhed, are, the effect produced by cupping glaffes applied to the furface of the body, viz. Temporary Inflammation, the fame event following the application of warm water or emollient poultices to particular parts; and the well known tendency of the blood to pufh towards the weakened part, wherever the coats of a veffel are divided. It remains however to be afcertained, whether increafe of action be compatible with a weakened ftate of the veffels. In order to judge of this, it is neceffary to recollect, that the parietes of the arteries are compofed of three coats, viz. a tough membranous one, commonly called elaftic, a mufcular one, and a fmooth thin membranous one; that thefe coats are fupplied with blood-veffels; and that the action of the arteries depends very much, though perhaps not folely, upon the influence of the nervous

* Vide quarto edition of Dr. M'Bride's Works, p. 159. et feq.

rous fyftem. Now the queftion is, Whether it be poffible for the mufcular coat to poffefs an increafed power, while the elaftic one is weakened? On theoretical principles this might be admitted: for the action of the elaftic coat is purely mechanical; while that of the mufcular one is, to a certain extent at leaft, influenced by the nervous fyftem.

The third opinion has been lately offered by Mr. Latta *. The principal argument in favour of it is deduced from the fwelling of the inflamed part; which, in his opinion, is owing to a partial ftagnation of blood: fo that there muft of confequence be a deficiency inftead of an increafe of action. The phenomena of inflammation, however, are by no means confiftent with this idea. The unufual heat of the part; the throbbing pain; and where thefe are confiderable, the accelerated action of the whole fanguiferous fyftem, clearly indicate an increafe of action in the veffels. This is farther proved by the confequences of inflammation; for mucous membranes, when inflamed, afford a greater than ufual quantity of their fecretions; diaphanous membranes pour out an increafed quantity of the thin fluid which they commonly fecrete, and in cellular parts, a fluid,

called

* Vid. A Practical Syftem of Surgery, by James Latta, Surgeon in Edinburgh, vol. i. pag. 9r.

called pus, is furnished by the veffels, differing in its properties from any of the natural fluids.

———

GENERAL PHENOMENA OF INFLAMMATION WITH-
IN THE HEAD, CHEST, AND BELLY.

ALL the contents of the great cavities above mentioned are compofed of three different mate-rials; viz. mucous membranes; what are called diaphanous membranes; and cellular and glandu-lar fubftance. Thefe materials exift either fingly or combined.

Firft, The inflammation of mucous membranes differs from that of other parts, in being attended with little pain. It is diftinguifhed by the fenfa-tion of increafed heat and of forenefs; and by the fecretion of the membrane being changed in its qualities and appearance. In the healthy ftate, the mucous membranes, viz. all thofe lining the internal cavities of the body to which the air has accefs, furnifh a mild tranfparent ropy fluid, like thin ftarch, quite infipid to the tafte, and poffef-fing all the qualities of mucilage. This fluid, when thofe membranes are infiamed, is of a very different nature. It is at firft thin and acrid, fre-quently excoriating the very membrane by which it is prepared; then its quantity is increafed, af-terwards it is gradually diminifhed, and becomes

2

thick

thick and opaque, refembling purulent matter; and finally, in favourable cafes, it returns by degrees to its natural ftate. This, however, is not the uniform courfe of thofe changes ; but it is unneceffiry to enumerate the various deviations here, as they are ftated under particular articles.

Ulcerations are the frequent confequences of this fpecies of inflammation, and in fome inftances alfo gangrene.

The changes produced on the membranes themfelves are an increafe of thicknefs and fenfibility, fucceeded, after the morbid affection has ceafed, by diminifhed energy and great fufceptibility of inflammation from the flighteft caufes.

Secondly, What are called diaphanous membranes are, the dura and pia mater, the pleura, the pericardium, the peritoneum, the tunica vaginalis teftis, the periofteum, and the capfular ligaments of the joints. The inflammation of thefe membranes is not marked by any peculiarity of fymptoms, unlefs it affect the general fyftem. It is then diftinguifhed moft commonly by acute pain of the affected part. This fpecies, however, is chiefly characterifed by the effects produced on the difeafed parts: for thefe membranes, when inflamed, are found thickened, opaque, and floughy, with a gelatinous or purulent exudation on their furface fometimes caufing preternatural adhefions. At other times, the cavities which they inveft are filled with a turbid ferum, with filaments floating in it. Sometimes this inflammation terminates in gangrene; but, except in fuch cafes, there never is any lofs of fubftance in the parts.

Thirdly, The inflammation of cellular and glandular parts, when feated externally, is very eafily diftinguifhed from every other fpecies. It is attended with throbbing pain, and it terminates moft frequently in fuppuration or abfcefs. But when it occurs in internal parts, it is not difcriminated with fuch facility. In general, it may be judged prefent, if, along with throbbing or fixed pain, increafed action of the whole vafcular fyftem take place, and more efpecially if, under thefe circumftances, blood drawn from a vein exhibit what is called a buffy coat.

The parts affected with this fpecies of inflammation, which is termed *Phlegmon*, are always fwelled, from the extravafation or effufion of ferum and lymph from the inflamed veffels, and alfo from the increafed quantity of blood in thofe veffels.

The termination of this fpecies is various. In fome cafes, the increafed action of the veffels ceafes, the effufed fluids, are abforbed, and the natural ftate of the part is reftored. This favourable event is ftyled *Refolution*. In other cafes, there is an effufion of a bland, opaque, yellowifh fluid, fomewhat like cream, called Pus. This is termed *Suppuration*. Sometimes, too, by fome morbid

morbid affection of the veffels, the red particles of the blood, as well as the lymph and ferum, are thrown out, and, inftead of pus, an acrid corrofive fluid is formed, which deftroys the neighbouring parts; and this termination is named *Gangrene.*

SECT. I. *PHRENITIS.*

PHRENZY occurs both as an idiopathic and a fymptomatic difeafe. The former fpecies feldom appears in this ifland, and the latter not very often.

The fymptoms of idiopathic phrenzy are: violent headach, attended with acute inflammatory fever; a rednefs of the face and eyes, an impatience of light or noife, a ftate of conftant watching, and the moft furious delirium *. The veffels of the head are turgid; the temporal arteries throb; the eyes fparkle, and are violently agitated; the tongue is dry, rough, and of a yellow or black colour. The patient is outrageous by fits: he grinds his teeth; his hands tremble; and he makes frequent violent attempts to get out of bed. Thefe fymptoms are preceded by long continued

B b 2   and

* Vide Cullen's Firft Lines, par. 291, et feq.

and almost conflant watching, or frightful dreams;
acute pains, at firſt in the neck and occiput, afterwards extending to the head; no defire for
drink; deep refpiration; irregular pulfe; fuppreffion of urine; and inability to recollect circumſtances that have juſt happened *.

Phrenitis is diftinguifhed from mania, by the
quicknefs of the pulfe; and from that fpecies of
delirium which occurs in low nervous fevers, and
is not produced by inflammation, by the appearance of the countenance and eyes. For, in true
phrenzy, the face is red, the features are rather
enlarged than ſhrunk, and the eyes protuberate
and fparkle; whereas, in the delirium fuperven-
ing to low fever, the face is pallid, the features
are ſhrunk, and the eyes pearly.

Although it has been fuppofed that fymptomatic phrenitis fupervenes to fevers of every kind †,
and to all inflammatory and painful complaints;
yet it is probable it only occurs in the latter
cafes.

The author of thefe remarks has often feen
phrenitis in lying-in women. He cannot however determine, whether it ought, under fuch
circumftances, to be ſtiled idiopathic or fymptomatic.

---

* Vid. Confpect. Therapiæ Specialis, auctore D. Joanne Juncke-
ro, pag. 520; from which the above defcription is chiefly taken.
† Vid. Sauvage Nofolog. Method. tom. i. pag. 458.

matic. In all the cafes which he has attended, the difeafe occurred within forty-eight hours after delivery; and in all of them too it proved fatal before the end of the fixth day. In fome of them the delivery had been natural, though tedious; while in others it had been exceedingly difficult. One of the patients had always had a ftrong pre-difpofition to inflammatory complaints, and more efpecially to inflammation of the parts within the thorax. The phrenzy was uniformly preceded by watchfulnefs, quick hard pulfe, and wildnefs in the eyes; and was ufhered in by violent pain of the head, great fufceptibility of the impreffion of light and noife, and a hurried mode of fpeaking. In every cafe the pupils of the eyes became quite dilated within forty-eight hours from the commencement of the phrenzy; fo that the light ceafed to make much impreffion. Throughout the courfe of the difeafe the pulfe varied from one hundred and twenty to one hundred and forty, and at laft it was intermitting and indiftinct. It was very remarkable, that, until a few hours before death, there appeared to a fuperficial obferver no morbid change in the features of the face; in fo much, that it was difficult to perfuade the attendants that any danger threatened. At that time, viz. fix or eight hours previous to the fatal event, a very fudden alteration happened. The outrageous delirium at once ceafed; the extremi-

ties

ties became cold; the features of the face fhrunk
amazingly; the eyes feemed fixed in their fockets,
and had a peculiar fhining appearance, as if covered with a glairy fluid. Partial fweats broke
out on the face, neck, and breaft; ftertorous
breathing gradually commenced; fubfultus tendinum fupervened: and the patient funk. This
difeafe was very accurately diftinguifhed from
puerperal fever, even at the beginning, by the
abfence of pain and forenefs in the abdomen, and
of uneafinefs of breathing.

Symptomatic phrenitis is known to threaten
in inflammatory or painful complaints, if, after
continued watching or frightful dreams, pain in
the head and tinnitus aurium take place, together
with a peculiar wildnefs in the appearance of the
eyes.

Phrenitis, whether idiopathic or fymptomatic,
is a very dangerous and alarming difeafe: for it
generally proves fatal between the third and feventh day, and if protracted beyond that time it
terminates in mania or fatuity. Amaurofis, too,
has fucceeded to an attack of phrenzy.

Sometimes however the difeafe ends favourably, by a critical difcharge of blood from fome
part of the body, moft commonly the nofe; or
by an univerfal fweat, or copious diarrhœa, or
even by depofition in the urine.

The unfavourable fymptoms generally enumerated

rated are : æruginous vomiting, the difcharge of white or grey coloured fæces, frequent attempts to fpit on the attendants, convulfions, and fleep not preceded by a critical difcharge. Subful us tendinum, or convulfions, and coma, with cold fweats and fluttering pulfe, announce the approach of death.

CASES of Symptomatic Phrenitis.

Case I. (VII. 13.)

A MAN, of a tall ftature and of a lean body, who had been fubject to inflammation of the cheft from dreffing flax, which was his trade, after having been fix or feven times affected with that inflammation, along with a vomiting of bilious green matter at one time, and delirium at another, had his voice at laft fo much injured, that he feemed to croak rather than to fpeak. On this account, he chofe out hemp which contained the leaft duft, and dreffed it in a feparate place from his companions. By which means he had juft recovered his natural voice, when, after having been fatigued with carrying a burden, he became affected with febrile coldnefs, and with a violent pain in the left breaft. With thefe fymptoms he was brought into the hofpital of Bologna. Previous to this he

had

had taken fome almond oil, and had been bled
from the left arm.   As he breathed with difficulty, and had no expectoration, blood was drawn
from the other arm.   He lay moft frequently on
the affected fide.   He vomited bilious matter of a
green colour.   On the fifth day he became delirious; being fometimes merry, fometimes melancholic, and fometimes fo furious as to fpit upon
thofe who approached him.   Blood was drawn
from his leg near the ancle; and a cataplafm,
compofed of new cheefe of the coarfeft kind mixed
with oil of violets, was applied to his head after
it had been fhaved, and was ordered to be renew-
ed three times a day.   In the mean time convulfive motions, at firft flight, and under the form of
fubfultus tendinum at the wrifts, afterwards more
confiderable, were obferved.   At length, neither
was his breathing difficult, nor did he complain
of any pain or uneafinefs; but, on the contrary,
when queftioned on that fubject, he always anfwered in the negative.   He fometimes however
cried out; and made water involuntarily, fo that
he wetted the bed.   His pulfe having become
weaker, though not irregular, he died about the
feventh day.

Appearances on Diffection.

ABDOMEN.   The ftomach was found; the pancreas was fomewhat hard, and rather thick.   The
edge of the liver fuperficially was livid to a confi-

I   derable

derable extent. The gall-bladder was contracted, and contained a small quantity of bile of a faint tobacco colour.

THORAX. No serous fluid was effused into the cavity of the thorax. The right lobe of the lungs was in a natural state, except that it was every where closely connected, by intervening membranes, both to the ribs and to the diaphragm. The left lobe was connected to the pleura only in a few places on the fore part. Its superior lobule, though in other respects found, was thin; and contained some purulent matter, of a white colour, in a kind of tubercle. These appearances, as well as the adhesions of the right lobe, seemed to be the effects of preceding inflammations. The inferior lobule was of a red colour; was hard, heavy, and of a compact substance; and contained in its superior part matter, or somewhat resembling matter, flowing through the branches of the bronchia. From these appearances present inflammation in the incipient stage of suppuration was indicated. The pleura also, on the same side, obviously appeared to be completely inflamed; for its blood-vessels were much more distinct than natural; and the whole membrane was drawn off from the ribs at a single pull. The diaphragm too, at that part of its centrum tendineum, which lies under the left lobe of the lungs, had its most minute vessels so much distended, that there could be no doubt of its being there inflamed. The pericardium contained fome turbid ferous fluid, of a reddifh colour. Through all the orifices of the heart polypous concretions projected. Thefe originated in the ventricles, and extended into the veffels. They were all of a folid fubftance, except the beginning of that which went into the pulmonary artery; for the part of it within the right ventricle, though very thick, was compofed of a yellow mucus-like fubftance.

HEAD. When the head was feparated from the trunk, long portions of coagulated blood were drawn out from the jugular veins, as fwords from their fcabbards. The veffels of the meninges were exceedingly diftended with blood. A polypous concretion, of a whitifh colour, and compact ftructure, not only filled the finus of the falx, but alfo extended into moft of the veins which communicate with that finus. Coagulated blood was alfo obferved in the other three finufes of the dura mater. The pia mater had all its veffels, even the moft minute, fo much diftended with blood, that it was all over of a very red colour. Beneath it, on the convolutions of the brain, fome ferous fluid was obferved. A fmall quantity of the fame kind of fluid, of a reddifh colour, was found in the lateral ventricles. Many hydatids, of a confiderable fize, were feen on the pofterior part of the plexus choroides. Not only were the veffels

on

on the furface of the lateral ventricles greatly diftended with blood, being much more diftinct than ufual; but alfo veffels, equally full and diftinct, appeared, on the flighteft incifion, through the corpora ftriata, the thalami nervorum opticorum, and throughout the whole medullary fubftance of the brain. In the cortical fubftance, however, both of the brain and cerebellum, the blood-veffels could fcarcely be feen.

Case II. (vii. 15.)

A woman, after having been for a confiderable time- in the hofpital of Padua, on account of a blow on the head, and after having been difmiffed cured, became affected with fever and delirium; which terminated in death.

Appearances on Diffection.

HEAD. There was no where any particular mark of the blow which fhe had formerly received. On the internal furface of the dura mater, florid fpots, like drops of blood, appeared. The veffels of the pia mater were diftended with blood. Serous fluid was found in fome places under that membrane, but not in the ventricles. Veficles were obferved in the pofterior part of the plexus choroides. A little yellow coloured matter was fituated at the fore part of the pineal gland. Every thing elfe was natural, except that the cere-

C c 2

bellum

bellum was very flabby. No polypous concretions were seen in any of the veffels.

CASE III. (VII. 2.)

A MAN, aged feventy years, by trade a potter,
naturally of a cheerful difpofition, much addicted
to drinking, after having undergone great fatigue
in bufinefs, and fuffered much uneafinefs of mind,
became affected with fever, attended with violent
pain in his left fide. He was received into the hofpital of Bologna. Blood was immediately drawn
from his left arm. On the fourth day, the febrile
fymptoms were much aggravated; and on the
fixth he grew fo delirious, that it was neceffary to
bind him down. His pulfe was quick, but not irregular; his refpiration was difficult; and he had
no expectoration. Blood was that day drawn
from his ancle. The fymptoms however continued to increafe in violence; fo that on the following day he had flertorous breathing, attended
with profufe fweat over his whole body, and foon
after died.

Appearances on Diffection.

EXTERNALLY. The face of the carcafe, and the
fuperior extremities, were of a yellow colour, as in
jaundiced perfons; the other parts of the body,
except the haunches, which were fomewhat livid,
were of the fame colour in a flighter degree.

ABDOMEN.

ABDOMEN. The omentum was obferved to be very fhort, and drawn upwards. The coat of the fpleen, on the lower part of the gibbous furface, was exceedingly hard. The liver was connected to the diaphragm by its whole convex furface, except the edge and parts immediately adjoining. The edge and the greateft part of the hollow furface were livid, to the depth of two lines. The remaining part of the liver was of a pale colour, and variegated like marble; and the whole of its fubftance was a little hard. The bile in the gallbladder was in fmall quantity, and refembled putrid blood, or water in which meat had been wafhed. Its paffage into the inteftines had not been obftructed; for the faeces were tinged with it. The inteftines were diftended with air. Yellow coloured fat adhered to their external furface, and a fmall quantity of ferous fluid furrounded them in the lower part of the pelvis. The portion of the inteftinal canal, and of the ureters which lay on the pelvis, was of a brown red colour. The urinary bladder was diftended with urine; and as well as its contents was yellow. The fanguiferous veffels towards the cervix, both pofteriorly and anteriorly, were turgid. On the right fide of the fundus, two fmall cells, each capable of containing a large cherry, and each communicating with the bladder by an opening, the diameter of which was as large as a lupin, were obferved. The parietes

rietes of thefe cells were of the fame ftructure as
the bladder itfelf. In the remaining parts of the
fundus, the beginnings of other cells of the fame
kind could be plainly perceived. A great number of enlarged and thickened veins furrounded
the right teftis on all fides; and the fubftance of
that body was fo compact that its ftructure could
not be developed as ufual. Below the teftis there
was a fmall offeous body. The tunica vaginalis
adhered to the right teftis everywhere, except on
the fuperior part, where there were two veficles
filled with a yellow coloured ferous fluid. The
fame coat adhered alfo to the other teftis, except
at the epidydimis, where there was a fpace filled
with fimilar fluid. On examining the penis, no
traces of fræium, except a fmall white mark, could
be difcovered. Nothing remarkable was feen in
the urethra, but a few minute granules of concreted mucus, like the powder of tobacco, fituated
on each fide of the feminal caruncles. Thefe bodies too appeared as if glued down to the urethra.
Throughout the proftate gland, efpecially on the
right fide, fimilar granules were found wherever
it was cut into.

As far as could be learned, this man had never
mentioned any complaint in the urinary fyftem.

THORAX. The firft circumftance which at-
tracted notice was the appearance of the cartilages
by which the inferior ribs are joined to the fternum

nium on the right fide; for they projecte l outwards, as if from fomething within. But, as nothing whatever was obferved internally which could account for the phenomenon, it is probable that it had been occafioned by the great exertion of the right pectoral mufcle during his youth, that had been perhaps neceffary in acquiring the rudiments of the trade by which the man had gained his livelihood. Every thing was found in the right cavity of the thorax; but in the left, a confiderable quantity of yellow coloured ferous fluid was found. When this was removed, fubftances, as if pieces of a thick yellow reticular and eafily lacerated membrane, appeared over the furface of the lungs. Thefe membranous portions were in greateft quantity on the inferior furface of the lower lobe and between the lobes. Almoft the whole of the inferior lobe was hard and heavy. On being cut into, its fubftance was found of an uncommon compact ftructure, more like that of the liver than of the lungs, and of a white colour. The inflammation feemed to have begun to terminate in fuppuration; for thick white matter was preffed out from feveral fmall orifices, probably openings into the bronchia. The upper part of the fuperior lobe was diftended with frothy blood, and was black and hard. The hardnefs appeared rather to have originated from fome former than from fome recent difeafe. The remaining

portion

portion was found. That lobe was connected to
the pleura laterally and anteriorly, by many ftrong
fafciculated fibres of a red colour, but apparently
of a membranous nature. It alfo adhered very
ftrongly to the pleura at the upper part of the
pleura; was there thickened, and could be eafily
pulled away from the ribs; which could be ftill
more readily done at the part where it lay under
the inferior lobe. At that place, it was more
thickened and of a deeper red. The external
furface of the pericardium, on the left fide, was of
a red colour, in confequence of the fmall fuperficial veffels being diftended with blood. Within
the pericardium there was a fmall quantity of a
yellow watery fluid. A thick polypous concretion was found in the right auricle; a fimilar fubftance, of a round form, was perceived in the pulmonary artery; and alfo one of the fame kind in
the aorta: and it appeared that, in the left auricle and ventricle, there were fome fimilar ones.
When thefe polypi were looked at longitudinally,
their fubftance feemed compofed partly of a foft
yellow mucus, and partly of a reddifh fibrous ftructure. On the internal furface of the aorta, above
the valves, there were offeous lamellæ. The an-
nular cartilages in the trunk of the afpera arteria,
and in the beginning of the bronchia, were offified in the middle part, by which they had a fmall
degree of flexibility only; and when the offified

I   portions

portions were broken, something like the rudiments of marrow appeared within them.

HEAD. A white, firm, slender polypous concretion was seen in the left lateral sinus, and also
in the fourth and third, and in some of the veins
communicating with the latter.   All the vessels of
the pia mater, on the left side of the head, and
even where it enveloped the cerebellum, were so
much distended with blood, that their trunks
were turgid, and their smallest branches were con-
spicuous.   There was a great quantity of serous
fluid on the external convolutions of the brain,
which shined through the pia mater, and resembled jelly.   The ventricles were not entirely des-
titute of fluid.   On the plexus choroides there
were hydatids ; but the plexuses were not discoloured. Neither the vessels which pass through
the medullary substance of the brain, nor those
which extend along the parietes of the lateral ventricles, appeared to be distended with blood, as the
vessels of the pia mater were.

CAUSES of Phrenitis.

PREDISPONENT CAUSE. As persons who are
liable to general inflammatory complaints, those
of a passionate disposition, and those addicted to
deep study, are observed to be principally subject

to phrenzy; it is evident that the predisponent cause is some morbid state of the vessels within the cranium.

EXCITING CAUSES. Every circumstance which tends to increase the action of the vessels within the head has been regarded as an exciting cause of phrenitis *. Such are, intoxication from the immoderate use of fermented liquors; blows on the cranium; and the exposure of the head for a considerable time to the influence of the solar rays. The operation of these causes needs no explanation. The process of parturition, when difficult, acts in the same way. The suppression of habitual evacuations is also an exciting cause.

In inflammatory affections of the parts within the thorax, the blood being prevented from passing readily through the lungs, cannot be returned as usual from the veins of the head; while, at the same time, the action of the arterial system is increased †: hence the cause of the frequent termination of peripneumony in phrenitis is obvious. ·

REMARKS ON THE CASES OF PHRENITIS.

THE first case illustrates the explanation offered respecting the cause of the termination of inflammatory

---

* Vide Dr. Cullen's First Lines, par. 294.

† Vide Van Swieten Comment. in Aphorism. Boerhaavi, No. CCLXXII.

matory complaints of the thorax in phrenitis ; for
the blood feemed accumulated in the jugulars, as
well as in the veins within the cranium. It is to
be obferved that, in this cafe, if the relation of
MORGAGNI can be credited, (and his general accuracy and fidelity cannot be doubted), both the
arteries and veins within the head were diftended
with blood.

In the fecond and third cafes, the ferous fluid on
the furface of the brain had proceeded from the
inflammation of the pia mater.

SECT. II.   *CYNANCHE TONSILLARIS;* OR,
COMMON INFLAMMATORY SORE THROAT.

THIS difeafe * confifts of inflammation of the
tonfils, the uvula, velum pendulum palati, and
mucous membrane lining the fauces, attended
with inflammatory fever.

The fymptoms are different according to the
degree and extent of the inflammation ; and, therefore, that an accurate defcription may be given,
it is neceffary to notice three varieties.

D d 2   1ft,

* Vid. Dr. Cullen's Firft Lines, par. 301, et feq. Home Princip. Med. p. 121. Quarin de curandis Febribus et Inflam. Comment. p. 223.

1*st*, In ordinary cafes one of the tonfils is firft affected; fo that deglutition, although painful, is not difficult. By degrees the inflammation extends to the other tonfil, to the uvula, to the velum pendulum palati, and along the membrane of the fauces; and is even communicated to one or both Euftachian tubes. The febrile fymptoms are then aggravated. Deglutition is very painful and difficult. Soft folids being more eafily fwallowed than liquids, the faliva is allowed to accumulate in the fauces, and excites a conftant hawking, attended with a degree of naufea. Pain is felt in the internal ear or ears; fometimes deafnefs is occafioned. The voice is hoarfe; and refpiration is fomewhat difficult. After fome days thefe fymptoms fubfide; and the inflammation terminates either by refolution or fuppuration. In the former cafe, falivation in various degrees takes place; and fometimes fmall floughy fpots of a white or yellow colour, and circumfcribed form, appear on the tonfils, or even over all the fauces. Thefe often continue for weeks after all the fymtoms of the difeafe have difappeared. In the latter, flight fhiverings are felt; the fwallowing becomes lefs difficult, and one or both tonfils when examined, appear no longer of a florid red colour, and exceffively diftended as they formerly were; but are foft, and white, or yellowifh, from containing pus. Previous, however, to this change,

the

the pain is fometimes in irritable habits fo great
as to induce convulfions.   The abfcefs at laft
burfts ; and its contents are either difcharged into
the œfophagus, which is known by a bitter tafte
being felt, or are hawked up mixed with a little
blood and mucus.   In fome cafes, inftead of a
fingle abfcefs there is a fucceffion of fmall abfceffes.   Soon after this, the patient regains his former good health.   But fometimes the tonfils remain indurated, and as it were fcirrhous, and
prove exceedingly troublefome.   In fome cafes,
too, the uvula is for many months relaxed, and
the tonfils are affected with a fpongy indolent fwelling.

2dly, In the fecond variety, the inflammatory
fymptoms both general and local, occur in a
much more violent degree.   The difeafe is ufhered in with fhivering ; then roughnefs of the throat
is felt, foon fucceeded by fwelling and inflammation of the tonfils, uvula, velum pendulum pala-.
ti, and whole mucous membrane of the fauces,
fo that the paffage to the nofe from the throat is
clofed up.   The root of the tongue, too, is fwelled and inflamed ; and alfo the mufcle that moves
the os hyoides.   The febrile fymptoms at the fame
time are violent ; the pulfe being often one hundred and forty.   Both deglutition and refpiration
are impeded.   The countenance of the patient is
                                              fwelled,

fwelled, and red; his eyes are fomewhat inflamed and prominent; and although he moves the
jaw with great difficulty, he is forced to keep his
mouth as open as poffible in order to breathe. He
complains of pain in the ears and head. Sometimes, at this period of the difeafe, epiftaxis takes
place, followed by mitigation of the fymptoms.
In other cafes, the fame event fucceeds a fwelling and rednefs of the fides of the neck. But
moft generally fmart rigors fupervene, and a large
abfcefs is formed in one or both tonfils. In fome
rare cafes, the matter being difcharged into the
larynx occafions fudden fuffocation; but it is commonly evacuated into the œfophagus, after which
the inflammatory fymptoms gradually fubfide.

3*dly*, Inftead of the inflammation being confined to the fauces, as in the two former fpecies, it
is fometimes extended downwards amongft the
mufcles of the larynx, and even along the membrane invefting that canal. Where the mufcles
alone are affected, there is a conftant fenfe of fuffocation from the difficulty of opening the glottis;
at the fame time, the voice is fharp and fhrill;
and great pain is felt in the act of fwallowing.
When the membrane invefting the larynx is inflamed, there is exceffive difficulty in breathing
and fpeaking, and the voice is ftridulous. In this
fpecies of the difeafe, the fymptoms proceed with
                                                    fuch

fuch rapidity, that the patient is within three or four days either relieved by the inflammation terminating in refolution in fome places, and in fuppuration in others ; or is carried off by fuffocation from the difcharge of matter into the trachea, or from the chink of the glottis being compleatly clofed up.

In the fecond and third varieties, phrenitis, or pneumonia, have been known to fupervene. The difeafe very feldom terminates in gangrene.

The fymptoms of inflammatory fever diftinguifh in general all the varieties of cynanche tonfillaris from every other fpecies of fore throat. In fome rare cafes, however, this criterion is not fufficient to difcriminate it at the beginning from the angina maligna. On fuch occafions, the appearance of efflorefcence on the furface of the body, the previous hiftory of the patient, and above all the nature of the prevailing epidemic, afford unequivocal marks of the difference of the two fpecies.

The inflammation in the firft variety is of a mixed nature; being feated in glandular bodies, and in a membrane which, while it partakes of the ftructure of mucous membranes, and of the cuticle, differs from both in poffeffing part of that mechanifm through which the fenfe of tafte is conveyed. The tonfils and uvula have therefore the phlegmonic fpecies of inflammation; while the membrane of the fauces is affected probably

with

with the eryfipelatous, altered fomewhat in its phenomena by the peculiarity of ſtructure of that membrane. The ſloughs ought perhaps to be con-ſidered either as ſuppurations of the mucous folli-cles, or of thoſe glandular bodies which ſerve to tranſmit the fenſe of taſte.

In the ſecond variety, the inflammation ſeems to differ from the former only in degree and ex-tent. It is indeed communicated to the root of the tongue, and the muſcles of the os hyoides; but this only aggravates the ſymptoms.

The inflammation of the third variety, however, is more complicated: for in it, as in the former caſes, not only are the glandular parts, the mem-brane of the fauces, and the muſcles of the os hy-oides, affected; but alſo the membrane lining the larynx, (which is a mucous one) and perhaps the ſubſtances which connect the cartilages of that organ.

CASE of Cynanche Tonsillaris.

(XLIV. 3.)

A CARPENTER, aged thirty-three years, of a tall ſtature, pretty corpulent, and of a large ſize, hav-ing, as far as could be learned, been in good health, returned home one evening during the cold ſea-

ſon

fon very much intoxicated, and greatly heated, both by the wine he had drank and by the fire at which he had fat. Having become very feverifh, and being affected with fore throat, a phyfician was fent for the fame evening, and he was bled in the arm. As the difeafe neverthelefs did not abate, he was in the morning brought into the hofpital at Padua, where venefection was repeated; but with fo little effect, that in the evening it was again had recourfe to, a quantity of blood being drawn from his foot. On the morning of the following day, the remedies, both external and internal, which had been already ufed, and which were ftill employed, having failed to produce any good effect, blood was again drawn from his arm, and at noon from the fublingual veins. The jugular could not be opened agreeably to the wifh of the phyficians, as the patient could not bear the pofture neceffary for that operation. Notwithftanding of all thefe means, the fymptoms of fever and reftlefsnefs were not only not moderated, but were even increafed; and at the fame time they were attended with difficulty of fwallowing, fpeaking, and breathing. On the following day, (which was the third of the diforder) he mentioned that he was affected with virulent gonorrhœa, and that he had been troubled with it for at leaft a fortnight. A vein in his foot was again opened on that day. The blood which had

hitherto been drawn had never had any cruſt on
its ſurface, but was ſomewhat hard, and contained a ſmall proportion of ſerum. His neck was a
little ſwelled, but not his face, which was not even
red. Within two hours after the laſt veneſection,
although the pulſe ſtill continued ſtrong, the patient himſelf was ſenſible of his approaching diſſo-
lution, which actually took place about noon of
the ſame day; though it happened in ſuch a manner, that it appeared to the attendants to have
been occaſioned by accident: for having called
for the gargle which he uſed, and having perhaps
put, unguardedly, ſome more of it into his mouth
than he intended, he inſtantly expired, as if he
had been ſuffocated by the fluid.

Appearances on Diſſection.

ABDOMEN. Nothing uncommon appeared within the abdominal cavity; except a globular body,
placed near the edge of the meſentery, which reſembled very much, in form, in colour, and in
ſize, one of the largeſt eggs that project from the
ovarium of a boiled hen. This was nothing elſe
than fat, of a more yellow colour than the reſt of
the fat of the body: it was included within a
ſingle membrane, in the form of a ſpherical bladder, having no intervening membranous lamellæ
as far as could be obſerved. The ſtate of the urethra was examined, in conſequence of the patient's having confeſſed himſelf affected with gonorrhœa.

norrhœa. The proftate gland would have appeared larger than ordinary, had not the penis alfo been of a large fize, as it generally is in a large body. The proftate, the caruncle, the veficulæ feminales, and every other part within the urethra, except the internal furface of that canal, which feemed to be fomewhat more moift and more red than ufual, were in a natural ftate. One of Couper's glands was wanting, a circumftance not very uncommon: the fubftance of the other was changed into a hard, firm, ligamentous-like body.

THORAX. The lungs were neither turgid nor inflamed, but were perfectly found. The membranes, however, which inveft thefe organs, like other membranes in this body, refifted the knife more than ufual. The left lobe of the lungs was clofely connected to the pleura, but the right was perfectly free. The pericardium contained a little red coloured fluid. This was not tinged in confequence of diffection; for, as it was congealed by the cold, it plainly appeared to be internally of a red colour. Within the heart, which was uncommonly large, being out of proportion to the body, (and it too was large) neither polypi nor any thing unufual were found. A fmall quantity of black blood only, which was neither too fluid nor too much coagulated, was feen within it. The aorta, from the valves at its origin quite to the celiac artery, exhibited many marks of difeafe; for

E e 2

here

here and there a few spots, which were not of a
bony hardness, were perceived on its internal surface.   Internally, too, except in the seat of thefe
spots, its surface was no where whitifh, but of a
red brown colour: neither was it smooth and
shining, as it generally is, but unequal, from small
and thin excrescences, of a red brown colour,
both externally and internally, and of various
forms and sizes.   The largeft of thefe might have
been covered with a lupin, the form of which it
much refembled.   When looked at, they appeared to be of a foft confiftence ; but when cut into,
they were found to be as compact and firm as the
parietes of the veffel.   This difeafed ftate was
much more confiderable in proportion as the artery approached the heart.   But it did not extend,
neither to the carotids, nor to the fubclavians, nor
even below the celiacs ; under which too the appearance of white spots became gradually lefs and
lefs.   The parietes of the artery were alfo harder
than ufual.   The fourth finus of Valfalva was obvioufly larger than natural, though not in a very
great degree.   The feptum of the auricles of the
heart had its furface next the pulmonary vein
marked with parallel furrows, which were not very
flight.

HEAD AND NECK.   The veffels of the brain,
both externally and internally, and not only within the ventricles, but alfo in different places with-

in

in the medullary fubſtance, were diſtended with
blood.   Thoſé veſſels which crept through the
left fide of the pia mater were more particularly
diſtended.   That membrane, like the other membranes of this body, when cut into, gave more than
uſual refiſtance.   The lateral ventricles contained
a ſmall quantity of bloody-like fluid.   The tongue
was thicker than uſual; the veſſels upon its ſuperior ſurface, from the baſis towards its apex, were
certainly enlarged in their diameters from the included blood.   The uvula and the velum pendulum
palati were found.   The membrane inveſting the
tonfils was confiderably thickened, as it contained
a quantity of yellow coloured ferum which refembled yellow jelly.   The tonfils themſelves were
ſwelled, and more eſpecially the left one, which
was harder than the other, and from which, when
preſſed upon or cut into, ſome pus was expelled.
The cartilages and muſcles proper to the larynx
were perfectly found; but the membrane inveſting that canal was difeaſed both externally and internally.   Internally it was a little redder than u-
ſual, as was alſo the contiguous part of the aſpera
arteria.   It was ſomewhat ſwelled too, but ſo
ſlightly that the chink of the glottis did not appear to be thereby ſtraitened.   The ſame membrane, where it covered the epiglottis, both on
the concave, convex, and lateral ſurfaces of that
organ, was ſwelled, and in ſome parts was of a flo-

ι   rid

rid red colour, in others of a brown red. Thefe appearances were lefs confpicuous on the concave furface than elfewhere, and on it were confined to the third part of its extent. When cut into, they were found to proceed from a collection of blood and ferum which diftended the membrane and the contiguous glandular bodies only. On the convex furface, thefe fluids feemed to have begun in part to be coverted into pus. Befides, the fame membrane externally, where it covered the larynx on the back part, together with the glandular bodies enveloped by it, were affected with confiderable inflammation, efpecially towards the fides. On each fide, the membrane projected in form of a protuberance nearly of the thicknefs of one's little finger. Thefe protuberances proceeding from the neighbourhood of the bafis of the cricoid cartilage, and converging in their afcent, reached fomewhat above the arytenoid cartilages. They were totally unconnected with thofe cartilages, and with the fuperior part of the larynx, although they adhered to the remaining and inferior part of that canal. Thefe bodies were like two inflamed condylomata, refembling in form and colour the appearance of the epiglottis already defcribed, except that they were of a deeper florid red colour, and had lefs of the brown taint. When diffected, they were found to confift of the membrane and its glands diftended with effufed blood and ferum.

The

The fwelling was moft confiderable on the left fide, which it has been remarked was the fide principally affected.

CAUSES of Cynanche Tonsillaris.

PREDISPONENT CAUSE. Befides the predifpofition to general inflammatory complaints, it appears that fome local affection of the parts predifpofes to inflammation of the throat; for it is obferved that fome perfons are wonderfully liable to it. Men it has been alledged are more fubject to this complaint than women. After the throat has been once inflamed, there exifts always a great predifpofition to a return of the inflammation.

EXCITING CAUSES. The exciting caufes commonly enumerated are, violent exercife of the organs of the voice, and the fudden or long continued application of cold to the throat, or to fome other part of the body. Hence the difeafe occurs moft frequently during viciffitudes from heat to cold in the ftate of the weather, at which time it is often epidemic. Common inflammatory fore throat is never communicated by contagion.

REMARKS

REMARKS on the Case of Cynanche Tonsil-
LARIS.

The preceding cafe is a well marked example
of the third variety of cynanche tonfillaris.

The feat of the difeafe is fo accurately pointed
out by the defcription of the appearances on diffection, that any additional obfervations are unneceffary.

Confidering the ftate of the veffels within the
cranium, it is a very remarkable circumftance that
delirium did not take place previous to death.

SECT. III. *PNEUMONIA;* or, Inflammation
of the Pleura and Lungs.

Inflammation of the pleura and lungs* is
indicated, if along with inflammatory fever there
be pain in fome part of the thorax attended with
cough, and difficulty of breathing.   The difeafe
is termed, Pneumonia, or Pleurify, Peripneumony, Pneumonic Inflammation, &c.

This

* A moft accurate defcription of this difeafe is contained in
Dr. Cullen's Firft Lines, p. 334. et feq.

This difeafe is commonly ufhered in by fhivering, fucceeded by the ufual fymptoms of inflammatory fever; a few hours after which pain in
fome part of the thorax, cough, and difficulty
of breathing fupervene. The patient, at the fame
time, has his face fwelled and livid, and is exceedingly reftlefs and anxious. If he have any
fhort flumbers, they are interrupted by frightful
dreams. The pulfe is generally frequent, full,
ftrong, and hard.

The pain is commonly fevere and pungent. It
is feated moft often in one part, viz. about the
middle of the fixth or feventh rib on either fide.
It moft frequently continues fixed.

The cough is more or lefs violent. It conftantly aggravates the pain. At the beginning it is
fometimes dry, but moft generally it is attended
with the expectoration of pellucid or frothy mucus.

The breathing is moft difficult and painful in
the act of infpiration. It is rendered more uneafy by pofture, particularly when lying on the
pained fide.

Such are the general fymptoms of the difeafe,
but a confiderable variety in this refpect is obferved. Thus, in fome cafes, it fteals on without being preceded by well marked figns of inflammatory fever; there having been no rigors
nor exceffive heat, and the pulfe being foft, and

Vol. I.   F f   fmall,

fmall, inftead of hard and full. The pain, too, is fometimes dull inftead of pungent, or the fenfation of oppreffive weight rather than of pain is felt.

Much variety in the ftate of the fymptoms is alfo obferved, even in cafes where the difeafe is clearly indicated at the beginning. Sometimes, for example, the pulfe is full and foft; at other times fmall and oppreffed; the pain extends from its original feat to other parts, as from the fide to the fcapula, or from the fternum to the clavicle; and the breathing is moft eafy when lying on the pained fide, or on the back.

After thefe fymptoms have continued in a greater or lefs degree for fome days, they are either mitigated or become rapidly much aggravated, or continue ftationary for a great many days.

In the firft cafe, hæmorrhagy from the nofe, or from the feat of piles takes place; or a profufe fweat breaks out over the whole body, or there is a copious expectoration of thick yellow matter tinged with blood, or a very great difcharge of loaded urine, or, it has been faid, a large evacuation of bilious matter by ftool. Sometimes two or more of thefe critical difcharges concur, as fweat and expectoration, or the hæmorrhagy and loaded urine. That fluid has in fome cafes a white or purulent appearance. Sometimes, too, the fymptoms

toms are mitigated by an eryfipelatous eruption in an external part of the body.

In the fecond cafe, no expectoration appears, or it is fuddenly checked, or only frothy mucus is coughed up. The difficulty of breathing, pain in the thorax, reftlefsnefs, and anxiety, continue to encreafe. The features fink, the pulfe becomes very quick, fmall, and even intermitting. Delirium fucceeded by coma, and fubfultus tendinum fupervenes. Partial fweats break forth, followed by coldnefs of the extremities; and death, or fudden fuffocation, happens. Sometimes too all thefe fymptoms are preceded by a temporary remiffion, which impofes on the patient and attendants.

In the third cafe, there is only a partial mitigation of the fymptoms, (or rather there is no increafe in their violence) until about the fourteenth day, when the pain ceafes; but the difficulty of breathing and quicknefs of the pulfe continue. Slight fhiverings, fucceeded by heat, foon after occur; and then the cough is found to be aggravated on the leaft motion, and the patient cannot lie on that fide which was formerly free from pain. Great debility and emaciation of the body enfue, which terminate in hectic fever and phthifis pulmonalis; or, after a great quantity of matter has been difcharged from the lungs, the patient gradually recovers good health; or, what often happens, he is fuddenly fuffocated. There is alfo in this cafe another variety: for after the fourteenth day, along with difficulty of breathing, increafed by the flighteft motion and cough, the patient is unable to lie on either fide; he feels a weighty fenfation above the diaphragm, and the noife of fluid within the cheft may be heard, or fluctuation even may be felt. Such cafes generally end in fuffocation.

Pneumonic inflammation occurs moft commonly during viciffitudes of the weather from heat to cold, and confequently is moft prevalent during fpring and autumn.

It attacks chiefly thofe in whom the inflammatory diathefis is ftrongly marked. No period of of life after puberty is exempt from it; but it moft generally occurs between the twentieth and the fortieth year.

The feat of this difeafe is commonly firft in the pleura, and from that communicated to the fubftance of the lungs. Moft authors before Dr. Cullen imagined that fometimes the inflammation was confined to the pleura lining the ribs, and fometimes it began in the parenchematous ftructure of the lungs, from whence it was extended to the pleura invefting thofe organs. The former of thefe cafes was ftyled Pleuritis, and was faid to be diftinguifhed from the latter (which was called Peripneumony) by the pulfe being very hard and tenfe, and the pain acute and pungent. But as
this

this diftinction did not feem to Dr. Cullen to be confirmed by the appearances on diffection, nor to be ufeful for practical purpofes, he wifhed it to be laid afide. If the reafons urged by him in fupport of his objection to the old divifion required any additional weight, the following hiftories would amply afford it: for there is not a fingle cafe, amidft the great number detailed, which can properly be called pleuritis; the lungs having been affected in every inftance.

When pneumonic inflammation terminates in the firft way defcribed, the phenomena clearly fhew that refolution had taken place, or that the effufed fluid had been evacuated.

The appearances on diffection, where it ends in the fecond manner, prove that either fuch a quantity of lymph and ferum, and fometimes even blood, had been effufed, as had choaked up the air cells of the lungs; or that the whole fubftance of the lungs had undergone fuch a change as to be indurated, and to have the cells much ftraitened, which is occafioned by a great number of minute veffels throughout the cellular ftructure being much diftended with blood.

The third termination depends upon fuppuraration within the fubftance of the lungs, or an exudation into the cavity of the cheft from the furface of the pleura having taken place. In the former cafe, the matter is fometimes deep-feated, and fometimes sometimes superficial. It is sometimes coughed up, at other times it is abforbed into the fyftem; but moft frequently it burfts fuddenly into the air cells, and fuffocates the patient. This conftitutes what is termed Vomica. In the latter cafe, which is named Empyema, the effufed fluid is in fome rare inftances re-abforbed; but moft generally it is in fuch quantity as to opprefs the lungs, by which the refpiration is more and more impeded, till at laft fuffocation enfues.

From thefe obfervations it is evident that inflammation of the pleura, like that of other organs compofed of cellular ftructure and diaphanous membranes, terminates in refolution, fuppuration, or exudation, fingly or combined. But it is to be remarked, that it has alfo a different termination from the inflammation of other parts: for the effufion which often precedes refolution as well as fuppuration fometimes proves fatal.

In fome rare cafes, there is an effufion of pure blood into the thorax, and in others the inflammation ends in gangrene *.

CASES

* The author of thefe remarks did not think it neceffary to give a more minute detail of pneumonic inflammation, and of the appearances on diffection, as both are fo very fully explained in the hiftories of the following cafes.

CASES of Pneumonic Inflammation.

Case I. (xx. 16.)

An unmarried woman, aged twenty years, was affected with pain, first in the left, and afterwards in the right side of the thorax, along with cough. She could not lie upon the right side. During the night she became delirious, and was affected with convulsions in such a manner that some of her limbs remained contracted. In this situation she died.

Appearances on Dissection.

ABDOMEN. The colon, very much distended with air, after having reached the stomach, proceeded in a straight line to below the middle of the belly; from whence it was immediately reflected towards the superior part of that cavity, and then followed its natural course. The appendicula vermiformis, of the thickness of a goose quill, and eight inches in length, lay in an oblique direction towards the right kidney. All the other viscera were in a natural state.

THORAX. The right lobe of the lungs was somewhat inflamed, especially at the posterior part, The left lobe, which everywhere adhered to the pleura lining the ribs and diaphragm, in such a

manner

manner that it could not be feparated without laceration, was very red. The heart contained no
polypous concretion; but in each ventricle black
clotted blood was found. It was alfo thick and
black in the veffels, but neverthelefs it was in a
fluid ftate.

HEAD. The brain was very found, though it
contained about its bafis a fmall quantity of ferous
fluid.

CASE II. (xx. 22.)

AN old man, of about fixty years of age, was
affected with pain in the right fide of the thorax.
He was alfo feverifh; and had a cough, attended
with expectoration. He lay with moft eafe upon
his back. The expectoration had become more
copious; but on the tenth day, after venefection
had been performed, it was fuppreffed. On the
eleventh day he died.

Appearances on Diffection.

THORAX. The left lobe of the lungs, although
it adhered everywhere to the pleura, was found.
The right lobe, on the contrary, although unconnected with the pleura, was on its pofterior part fo
much inflamed as to refemble in its fubftance folid
flefh. Two polypous concretions, nearly fimilar
to each other, one in each ventricle, were found
in the heart.

2   CASE

CASE III. (xx. 24.)

A MAN, of about forty years of age, who had long laboured under two herniæ, became affected with a pain in the thorax, and difficult refpiration. At the beginning of the difeafe he lay moft eafily on the right fide, and after that on the left; but he could by no means lie upon the back. He expectorated much matter. He was obliged to hold his neck erect, in order to breathe. On the tenth day he died.

Appearances on Diffection.

ABDOMEN. The cavity of the tunica vaginalis teftis was filled with ferous fluid; and from that coat itfelf fome drops of ferum could be preffed out. This conftituted, on one fide, a hydrocele. Upon the other fide there was a varicofe production of the veins of the fcrotum, which reprefented pretty exactly the figure of a chain. Thus what appeared to be herniæ, were in fact a hydrocele and cirfocele.

THORAX. Both lobes of the lungs adhered to the pleura; in fuch a manner, however, that they could be feparated without laceration. At the parts next the back they were inflamed, and much indurated. The pericardium contained no fluid. In the heart three polypous concretions were obferved. One of thefe was in the left ventricle:

VOL. I.   G g   the

the remaining two were thicker, and of a firmer
fubftance; and were placed, one in the right
ventricle, and the other in the correfponding auricle.

C A S E IV. (xx. 26.)

A MAN, aged fifty years, who had from his
birth been of a brownifh complexion and atrabi-
lious temperament, and who had been much addicted to venery, became affected with difficult
refpiration, like what is occafioned by catarrh.
His difeafe having gradually increafed in violence,
he was in about a month compelled to confine
himfelf to bed. His voice was fhrill and loud.
He had a copious expectoration of watery matter.
His refpiration was exceedingly difficult, attended
with a fenfe of ftrangulation in the throat; and
he lay with great difficulty on either fide, but with
moft difficulty on the left, his breathing being then
more uneafy. His refpiration at laft became pant-
ing; nor could he breathe unlefs his neck was
erect. In the progrefs of the difeafe the copious
watery expectoration was changed into a fmall
quantity of thick, vifcid, ftinking matter, and his
refpiration grew eafier; but the fenfe of fuffocation ftill remained. The fever was always flight.
At length he died.

Appearances

Appearances on Diffection.

ABDOMEN. The fpleen was found everywhere connected to the neighbouring parts by intervening membranes: its arteries were in a cartilaginous ftate. In the right fide of the fcrotum a hydrocele was obferved. This was occafioned by a quantity of watery fluid which was collected between the tunica albuginea and the teftis. When that membrane was compreffed, efpecially at the fide of the large blood-veffels, fome limpid drops of fluid flowed out. When this fluid was expofed to a ftrong heat, it became coagulated into a very white fubftance: when placed over a gentle fire, it was gradually evaporated, and left behind it marks of lymph.

THORAX. The right lobe of the lungs differed little from its natural ftate. The left adhered at every point to the pleura: its fubftance was indurated and inflamed. In each ventricle of the heart a polypous concretion was feen; and, contrary to the obfervation of Valfalva, that in the left ventricle was the largeft. It was not allowed to examine the fauces, where perhaps the principal difeafe lay.

CASE V. (XX. 55.)

A MAN, aged forty years, of a lean habit of body, who had begun nearly two years before to

complain

complain of a pain, more or lefs violent at different times, about his ftomach, together with hardnefs at that part, fometimes attended with a fimple diarrhœa; and who had lately been affected with a flow continued fever, which lafted for fix or feven months, and reduced him much, was feized with an acute pricking pain in the left part of the thorax. He lay with great difficulty on that fide, though he could eafily lie on the other. He had cough, and at the beginning a little expectoration. The expectoration having ceafed, he died on the tenth day after the firft attack of the pain in the thorax.

Appearances on Diffection.

THORAX. The fuperior portion of the left lobe of the lungs had entirely degenerated into a hard tumor, compreffing on all fides the neighbouring parts, and connected with the pleura in fome places by fmall filaments. The pleura itfelf was inflamed. The right lobe, although it adhered fo ftrongly to the pleura that it could fcarcely be feparated without laceration, was perfectly found. A moderate quantity of watery fluid was found within the pericardium. The ventricles of the heart contained polypous concretions. That in the left, which was very fmall, extended into the aorta; that in the right, which was larger, was continued into the pulmonary artery.

CASE

Case VI. (xxi. 4.)

An old man, feventy four years of age, of a low ftature of body, who ufed to go about victualling houfes; after having been for feveral years fubject to inflammation of the lungs, at laft died in the hofpital of Padua, on the eighth day, of a difeafe of that kind. The particular fymptoms of the cafe could not be accurately learned; but there was no doubt that no mark of difeafed heart, fuch as deliquia, palpitations, or irregularities of the pulfe, had been obferved.

Appearances on Diffection.

Thorax. The lungs were univerfally connected with the pleura. The upper part of the right lobe was indurated, and of a blackifh colour. The fanguiferous veffels of the fame lobe were very clofely connected with the bronchia; and one of thefe veffels was much dilated to the extent of fome fingers breadth, after which it became of its ordinary fize. The trunk of the bronchial artery, where it arifes from the aorta, was nearly three times larger than ufual. A tubercle, of the fize and form of an ordinary cherry, appeared on the pofterior part of the left ventricle of the heart, at the diftance of two fingers breadth from the apex. One half of this tubercle projected above the fur-face, and the other half was buried within the

fubftance

fubftance of the heart. It refembled one of thofe hydatids which are feen in other organs, as the lungs or kidneys, and are fo placed as to jut out-wards. When punctured with the knife, it dif-charged a little watery fluid, but a more turbid fluid remained. When the tubercle was fully o-pened, this was forced out along with a fmall membrane, in which there were fome white mu-cous fubftances, together with a particle of a ten-dinous hardnefs. This fmall membrane feemed to ferve as the inner coat of the tubercle; for, exter-nally another coat, which was thick and white, and rough and unequal on the infide, furrounded it entirely. This coat was quite found, as were alfo the parts adjoining to the tubercle. The left auricle of the heart was much longer than ufual. On the internal furface of the aorta, efpecially at its arch, and near the heart, numerous bony valves were obferved; but, behind the femilunar valves the beginnings only of fuch fcales appeared. Some more of thefe fcales were feen in other parts of that veffel, even as far as the origin of the cœliac.

C A S E VII. (LXX. 10.)

An old woman having been affected with inflam-mation of the lungs, together with diarrhœa, died.

Appearances on Diffection.

ABDOMEN. Each ovarium, and more efpecially

ly the right one, was of a white colour, and very
much enlarged; and appeared of a knotty texture, from tubercles or hydatids, fome of which
were larger than the others, and were filled with
watery fluid, which fpouted out from them when
they were cut into. The Fallopian tubes were
totally unconnected with the ovaria, and were in
a natural ftate. The internal furface of the fundus uteri was of a red colour, inclining to black,
in confequence of blood which lay under it in
feveral places. When the uterus was externally preffed upon, in order to force out that blood,
the internal furface was found to be fo flaccid that
it was torn. When the pofterior crural nerve was
feparated into two portions, in a longitudinal direction, which was done by the finger alone, a
veffel of the diameter of the twelfth part of an
inch, filled with blood, appeared amongft its fibres,
almoft in a direction parallel to them, fituated
near the axis of the nerve. The internal coat of
the urethra was quite black, from its veffels (which
lay longitudinally on it, and parallel to each o-
ther) being much diftended. This blacknefs became greater towards the meatus; that part was
flabby, and allowed a relaxed portion of the internal membrane to be prolapfed without it on
the right fide. The bladder was in a found ftate:
the two protuberating bodies which proceed from

the

the ureters, appeared meeting nearly at an angle, at the diftance of a finger's breadth from the orifice of the bladder. No mark of the roundifh body, called uvula, could be diftinguifhed, neither about the orifice of the bladder, nor in the fpace between it and the infertion of the ureters, nor in the contiguous part of the urethra.

THORAX. The inferior lobe of the lungs, on the right fide, which was very large and heavy, adhered ftrongly to the pleura. When cut into, its fubftance was found to refemble that of a boiled liver. The heart was very large in proportion to the ftature of the woman, which was of a moderate fize. This was not owing to the thinnefs or extenfion of its parietes, for they were of the ordinary thicknefs, or rather were thicker than ufual. The columnæ, as well as the fafciculi of the ventricles, and the fafciculi of the auricles, were uncommonly thick. The orifices of the coronary arteries exceeded the fixth part of an inch in diameter; and the artery which lay next them on one fide, called arteria adipofa, had alfo an orifice which was not very fmall. Within the left auricle, befides feveral moderate fized mouths of velfels, one as large as that of the coronary arteries juft ftated, appeared open. On cutting into the veffel to which it belonged, it was obferved that two or three fmall veins returning the blood from the parietes of the auricle, ran into it.

s   The

The ventricles were full of polypous concretions.
The aorta on the right fide, where it begins to
defcend, had its internal furface unequal, as it
projected inwardly, and contained between its
coats a fubftance of a bony hardnefs, which might
have been covered by the nail of one's thumb.
Below the two upper pair of the inferior intercoftal arteries, inftead of the three fucceffive pairs
which fhould have appeared, a fingle veffel was
at each of the places fent off from the middle of
the aorta. Thefe were neither of a larger fize
than ufual, nor were they immediately divided
into two branches, which commonly is the cafe,
when one veffel is fent off inftead of a pair; for
each extended as a fingle trunk for the fpace of a
finger's breadth at leaft. The diaphragm had inftead of one foramen, for allowing the tranfmiffion
of the vena cava, two foramina divided from each
other by a thin partition.

Case VIII. (xxi. 35.)

A man, of a middle age, and healthy habit
of body, rather lean than fat, by trade a ftone-
cutter, having been affected firft with fever, and
immediately after with acute pricking pain in the
left fide, and fuch a loofe ftate of the bowels that
he had within twenty-four hours eight fluid ftools,
which were not of a yellowifh colour, nor difcharg-

ed with any uneafinefs, fucceeded by an oppreffive pain in the thorax, was on the fifth day of the difeafe brought into the hofpital of Bologna, having had no affiftance while in his own houfe. He was bled; and the blood drawn was exceedingly thick, but exhibited no buffy cruft. Along with the fymptoms already enumerated, which continued to the laft, he expectorated fome matter tinged with blood, but the expectoration was not of long duration. At length, on the eleventh day, he grew confufed in his mind, and fomewhat delirious, and his pulfe, which before had been tenfe and not intermitting, having become imperceptible, he died.

Appearances on Diffection.

ABDOMEN. The colon, being fomewhat diftended with air, appeared immediately on opening the belly. From the middle of that portion, which is generally extended acrofs the belly, it turned downwards to the umbilicus; from whence it again mounted upwards, but not very high: in other refpects it was in a natural ftate. Some parts of the fmall inteftines appeared as if inflamed. The liver was, both externally and internally, of a pale colour. The gall-bladder was almoft empty, as it contained but a very few drops of bile, that tinged paper, over which it was fpread, of a dirty yellow brown colour. The fpleen was larger and more flabby than ufual.

THORAX.

THORAX. The cheſt contained no fluid. Both
lobes of the lungs were on the anterior part turgid, and were there both of a natural colour and
conſiſtence; but, on the lower and poſterior ſurface, and in the left ſide on the whole lateral ſurface, they were cloſely connected to the parietes
of the thorax. The pleura, wherever the lungs ·
adhered, was unequal in its ſurface: towards the
back it was blackiſh, and on the left ſide was
thickened, and was of an unnatural colour. The
diaphragm, both on its muſcular and tendinous
parts to which the lungs had adhered, was of a
reddiſh brown colour, and had its blood-veſſels
much more diſtinct than common. The lungs
were very heavy: they were black on their whole
poſterior and lower ſurface, and the blackneſs penetrated their internal ſtructure, which was compact and ſomewhat hardened. On the left ſide a
conſiderable portion of their ſubſtance was found
to be more indurated and more compact; while
on the anterior part, eſpecially on the right ſide,
it was of a ſoft thin conſiſtence, and when cut into appeared of a roſe colour. The pericardium
contained a little more fluid than uſual. The veſſels on the ſurface of the heart, eſpecially on the
flat ſurface, were turgid with blood. Both the
vena cava and the right auricle were alſo diſtended with blood. The blood was exceedingly black;
and contained no polypous concretions, (and none

appeared

appeared, neither in the ventricles of the heart,
nor in any of the veffels) nor almoft any coagulated fubftance of any kind.

-        C A S E IX. (xx. 51.)

A YOUNG man above twenty years of age, who
had previoufly been affected with chronic fever,
was feized with acute fever, attended with pain
of the breaft, difficulty of breathing, and the expectoration of a fmall quantity of matter tinged
with the colour of blood. During the firft days of
the difeafe he lay for the moft part on his left fide;
and during the latter days he lay conftantly on
that fide. About the fixteenth day he died.

Appearances on Diffection.

ABDOMEN. The fpleen was three times larger
than ufual.

THORAX. The left lobe of the lungs was fo
much fwelled, that it filled entirely the cavity in
which it was placed. It was univerfally indurated and inflamed, and adhered everywhere to the
pleura. That membrane being of a reddifh colour, exhibited marks of incipient inflammation.
The pericardium was not only filled, but alfo much
diftended, with a fluid like cow-milk whey, which
had depofited fome concretions on the furface of
the heart. A polypous concretion, of a very fmall
fize, and very flaccid, was found in each ventricle
of the heart: that in the right was the largeft.

Branches

Branches of thefe polypi extended into the arteries, auricles, and veins: thofe in the veins were the larger.

CASE X. (XXI. 19.)

A BUTCHER, aged feventy-eight years, of a tall ftature and of a fallow complexion, who at other times had been troubled with the expectoration of bloody matter, had for four days felt an internal acute pricking pain a little below the left breaft; when he was, on that account, admitted into the hofpital of Bologna. His pulfe was very irregular, intermitted often, was quick and feeble. He had frequent cough, attended with a found nearly refembling the barking of a dog. The matter which he expectorated was thick, and contained polypous-like fubftances of a white colour. His refpiration was difficult; and he could only lie on his back. Eight ounces of blood were drawn on the day of his admiffion into the hofpital, namely, the fifth day of the difeafe. A yellow cruft, two inches deep, of a very firm confiftence, and marked with livid fpots on its external furface which was hollow, appeared on the top of the blood. The craffamentum below the cruft was melted down into a number of little lumps. It contained no more than a fingle fpoonful of ferum, which was turbid. Blood afterwards was feen

feen on the expectorated matter. The fymptoms
not having been in the leaft alleviated, venefection
was again performed on the feventh day. The cruft
of the blood then drawn was thin; the ferum, which
was of a golden colour, was in proper proportion,
and the craffamentum, was of the natural confiftence. On the eighth day, the expectoration having become diminifhed, and the other fymptoms
having continued, he could no longer fpeak, but
turned himfelf on the right fide, in which fituation
he died in a placid manner, without any ftertor.

Appearances on Diffection.

ABDOMEN. The abdomen was raifed into a
flaccid and flightly livid tumour, at the right ileum;
this was found to proceed from the colon being
at that part much diftended with air. Another
tumour, of a fmall fize, appeared in the left groin.
It confifted of the inguinal gland, which meafured
one inch in thicknefs, one and a half in breadth,
and two in length. When cut into, the greateft
part of its fubftance had a natural appearance;
the remainder, however, was of a whitifh colour,
and feemed to confift of fmall round particles.
The fcrotum was fwelled on both fides, but efpecially on the left; and in the lower part of the
right fide was of a red colour. It contained three
fwellings. The firft of thefe appeared on the left
fide, at the fuperior part of the tefticle, and confifted of a facculus, extending from the cavity of
the

the abdomen, and containing nothing but a long fold of the omentum, which was rugous, and could be eafily drawn forward. Not only the rugous ap-pearance, but alfo the uncommon convolutions of the ileum more efpecially, and of the colon too in fome refpects, obferved in the belly, clearly proved that this facculus had been at other times diftend-ed with a prolapfed inteftine. What conftituted the other two fwellings, was a collection of yel-low watery fluid within the tunica vaginalis teftis on each fide. That coat was thickened. This fluid was fuppofed to have been effufed from a ruptured hydatid; and a fmall thick veficle, al-ready almoft folid, and of a flefhy colour, append-ing by a peduncle from the tunica albuginea, where it invefts the body of the tefticle near the large bulb of the epidydimis, feemed to form the remains of that hydatid. The veficulæ feminales, and more efpecially the veffels which return the blood from them, were larger than ufual; but they ap-peared to be in an incipient varicofe ftate, in con-fequence of the fwellings already defcribed, rather than to have been enlarged from the falacious dif-pofition of the man: for the prepuce, without any mark of previous difeafe, was contracted before the glans, juft as it is in young men who have never had any connection with women.

THORAX. The right lobe of the lungs was eve-ry where connected ftrongly to the pleura, from

which

which however it could be eafily feparated without laceration, except at its fuperior part. At
that part it was blended with the pleura, and
within its fubftance which was of a livid black
colour, it contained a number of fmall round cells,
each furrounded with its own proper covering, internally fmooth, and almoft empty, and having
no outlet as far as could be obferved. Thefe appeared to be the marks of an old difeafe. When
this part of the lungs was cut into, a fmell refembling that of four whey, or of the breath of
children affected with worms, was perceived. The
upper part of the left lobe was not difeafed, except that it contained too great a proportion of
ferous fluid. But its lower portion was heavy,
red, and indurated like the fubftance of the liver.
It adhered here and there to the pleura, and in
different places was covered with broad white
fubftances like mucous membranes. The furface
of the diaphragm next the lobe was covered with
a fimilar membrane, only that it was red, and
lay under a little turbid ferum. Both thefe membranes were readily feparated without injuring the
furface of the lungs or diaphragm. The pleura
invefting the parietes of the thorax and the diaphragm, alfo, was not only of the hardnefs of cartilage or bone, but even confifted here and there
of thin bony laminæ. From its internal furface,
and efpecially at the parts occupied by the bony

I   laminæ,

laminæ, numerous offeous, tubercular bodies, of a hemifpherical form, and of the fize of a vetch, pro-jected. All thefe appearances were more ftrongly marked on the right, than on the left fide. In no part was the pleura red or inflamed. Within the pericardium a moderate quantity of reddifh coloured watery fluid was obferved. The right ventricle of the heart contained a large fmooth polypous concretion of a flefhy colour, extending into the neighbouring auricle; and alfo one of a round form and thicker confiftence, which went into the pulmonary artery. A fimilar polypus was feen ftretching into the aorta from the left ventricle; and a very fmall one, not unlike an incruftation, lay in the left auricle. In both ventricles of the heart, and in the aorta, there was alfo fome black half-coagulated blood.

Case XI. (xxi. 2.)

A woman affected with peripneumony, attend-ed with irregularity of the pulfe, died in the hof-pital of Bologna.

Appearances on Diſſection.

Abdomen. In the gall-bladder, there were two black calculi, of a pretty hard confiftence, approaching to the form of a cube, and of an un-equal fize, but neither of them were fmall.

Thorax. The fubftance of the lungs was as

compact as that of the liver. The pericardium
contained a large quantity of fluid. The greatest
part of the external surface of the heart appeared
at first fight to be corroded; but that was not the
cafe; for irregular concretions adhering to it
formed that appearance. Thefe were readily removed; and then the surface of the heart was
found quite smooth and found. To the internal
surface of the pericardium, other concretions of
the same nature, but totally unconnected with
the former, adhered. It was therefore probable,
that all thefe concretions were formed from the
fluid contained in the pericardium. Large white
polypous concretions, which were not of a very
firm confistence, were found in the ventricles of the
heart and in the right auricle.

C A S E XII. (xx. 3.)

A man, aged forty years, who had, on account
of a flight wound in his leg, been for a considerable time in the hofpital at Bologna, where he had
lain in the fame bed with a man who was carried off by the difcharge from a large abfcefs in
his thigh, became affected with acute fever; attended with cough, the expectoration of matter
first tinged with blood, and afterwards of a greenifh colour, difficulty of breathing, and pain in
the right fide. Under thefe fymptoms, while lying

ing on his right fide (on which fide he lay eafily), he died about the fourteenth day from the beginning of the pneumonic affection.

Appearances on Diffection.

THORAX. The left lobe of the lungs was found and unconnected with the contiguous parts. The upper lobule and the fuperior part of the adjoining lobule on the right fide, were towards the back, fwelled, inflamed, and much indurated, and adhered clofely to the pleura by membranous bands. That membrane however, exhibited no marks of inflammation. Each ventricle of the heart contained a polypous concretion of no inconfiderable fize; that in the right was the largeft. Although thefe concretions extended not only into the auricles and veins, but alfo into the arteries, they were not continued fo far within the latter veffels as within the former.

CASE XIII. (XXI. 33.)

A SOLDIER, of a middle age, rather of a lean habit of body, was received into the hofpital of Padua, in confequence of being affected with an acute pricking pain on one fide, together with fever, cough, and difficult refpiration. Thefe complaints continued for fome days; and then having become almoft comatofe, being affected with flight delirium and tremor in his hands, and .

his pulfe having at laft been imperceptible, he died.

Appearances on Diffection.

ABDOMEN. The omentum adhered to the beginning of the colon. When it was removed, that inteftine appeared in a very unufual fituation: for, after having reached the liver, which it fcarcely touched, it extended immediately from thence below the umbilical region, and then was reflected upwards into the left hypochondrium. All that portion was much dilated with air. The pancreas was fomewhat indurated. The fpleen was confiderably larger than ufual, and on that account extended lower down. The liver was alfo large, and over the greateft part of a white colour, and did not appear quite found when cut into, efpecially on the right fide. The ductus communis choledochus was wider than ufual.

THORAX. The lungs adhered almoft everywhere, except at the anterior furface, where they were nearly of the natural colour, very clofely to the pleura. The right lobe was much enlarged, was heavy, and throughout indurated, except in a fmall portion on the fuperior part, and on the anterior furface. It had a fmooth and uniform furface like that of the liver, and internally it exhibited no reticulated vafcular appearance, but refembled the fubftance of the liver when boiled. The left lobe, when cut into, appeared fomewhat

harder

harder than ufual, and was of a blackifh red colour; in other refpects it was in a natural ftate.
The pleura on the left fide was colourlefs, or rather whitifh ; but, on almoft the whole of the
right fide, it appeared of a red colour, obvioufly
from inflammation. The cheft contained no extravafated fluid. When the lungs were cut off
from the afpera arteria, which in this man was remarkably wide, and furnifhed about its divifion
with larger and more numerous bronchial glands
than ufual, a large quantity of bloody frothy watery fluid immediately flowed out from the right
lobe of the lungs, and a fmall quantity alfo from
the left. In the pericardium, there was a fmall
quantity of yellow fluid very flightly tinged with
red. Within both ventricles of the heart white
polypous concretions, with black grumous blood
everywhere adhering to them, were found. They
extended for a confiderable way into the large
veffels. Similar concretions, refembling tape-
worms, had been found in the iliacs ; and fmaller
ones, like afcarides, appeared on the longitudinal
finus of the head.

HEAD. A confiderable quantity of watery fluid
was difcharged when the cranium was opened.
The greateft part of the veffels of the pia mater
were turgid with blood. When the medullary
fubftance of the brain was cut into, a great many
drops of blood, fome of them large and fome of
them

them fmall, appeared here and there. A large
quantity of turbid watery fluid was found in the
lateral ventricles. On the choroid plexufes, which
were not pale, many veficles, fome of them of a
pretty large fize and filled with fluid, were feen.
The fornix and medulla oblongata were flabby.

Case XIV. (xx. 11.)

An unmarried woman, aged fifty-five years, became affected with pain in the thorax, efpecially
about the fternum, attended with fever. At the
fame time, fhe complained of pain in her head,
had difficulty of breathing, and expectorated a
large quantity of matter. Diarrhœa having fupervened, all the fymptoms having been much
aggravated, and the difficulty of refpiration having increafed to fuch a degree that fhe could not
breathe but in the erect pofture, fhe died about
the ninth day of the difeafe.

Appearances on Diffection.

Abdomen. The gall-bladder was much diftended with bile. Both ovaria were greatly indurated.

Thorax. The left lobe of the lungs adhered
fo firmly to the pleura lining the ribs, that it
could not be feparated without laceration. The
right lobe, having adhered to the mediaftinum
in its whole extent, and to the pleura invefting
the

the fuperior ribs, was torn, in confequence of its putridity, when an attempt was made to feparate it from its connections; and at the fame time dif-charged, from an abfcefs which it contained, a large quantity of fanious matter of a pale red co-lour. The pericardium was filled with ferous fluid. In the right ventricle of the heart a large poly-pous concretion, which filled the adjoining auri-cle, and was extended beyond the orifices of the veins communicating with it, was found. No po-lypi were obferved in the left ventricle; but a fmall one was feen at the mouth of the aorta, and an incipient one in the pulmonary vein. The blood, it is to be obferved, in this body, had loft its fluidity.

Case XV. (x. 5.)

An old man, aged fixty-five years, who had two years before been affected with palfy in the right arm, and after having been freed from that difeafe had been accuftomed to complain frequent-ly of a pain in his head, was brought into the hofpital at Bologna on account of inflammation of the lungs. When admitted, he complained much of a pain and fenfation of weight in the left fide of the thorax, and he expectorated purulent mat-ter. The expectoration having ceafed for many hours,

hours, he fuddenly died on the twenty-third day of the difeafe.

Appearances on Diſſection.

THORAX. The left lobe of the lungs, which was fwelled and indurated, filled up the whole cavity on that fide; and contained an ulcer, in which a large quantity of fanies was collected, and that matter was diffufed through the whole fubftance of that part of the lungs. The pleura was perfectly found. In the right fide of the heart there was a large polypous concretion, which was extended near a foot and a half within the vena cava, and followed the divifions of that veffel. In the left fide of the heart another polypus was feen, but it fcarcely reached the orifice of the aorta.

HEAD. The quantity of half a pint of watery fluid was found within the ventricles of the brain. The glandules of the plexus choroides were turgid. No other preternatural appearance occurred.

CASE XVI. (XXI. 45.)

A MAN, who had been affected with pneumonic inflammation in a very violent degree, was rapidly carried off.

Appearances on Diſſection.

THORAX. The lungs were fo much fwelled

I   that

that they filled the whole cavity of the thorax.
The left lobe, in which fide the pain had been
feated, appeared wholly inflamed, and of a black
colour; and befides, towards the lower parts, was
ftuffed up with a whitifh ichor, like that of an abfcefs. The pleura was everywhere inflamed, livid,
and marked with bloody-like points. The intercoftal mufcles, and particularly the internal ones
on the left fide, were inflamed to fuch a degree
that they appeared bruifed. Coagula of black
blood, not unlike polypous concretions, were found
in the ventricles of the heart, and in the contiguous large veffels, both arteries and veins.

Case XVII. (xx. 9.)

A lady, aged forty-five years, who was very
fat, and feemed to have a habit abounding with
ferum, in the beginning of the ninth month of
pregnancy, after having unguardedly expofed herfelf to a very cold air, was affected with univerfal
fhivering, fucceeded by confiderable heat, great
thirft, difficulty of breathing, very troublefome
cough, a moft oppreffive pain in the right fide of
the thorax extending to the fcapula, and bilious
vomiting. At the fame time, her pulfe was frequent, hard, and quick in the contraction of the
artery; and fhe had great reftlefsnefs and toffing
of the whole body, a fymptom which continued

VOL. I.   K k   exceedingly

exceedingly troublefome during the fucceeding days. The vomiting feemed to afford a little re-lief.

This patient, when in health, had always had difficulty of breathing, efpecially after motion. She had alfo been often troubled with flight cough, attended with copious expectoration ; and almoft every day fhe had been accuftomed to vomit in the morning (after having paffed a reftlefs night) a thick vifcid matter, after which thofe complaints feemed to be relieved. She had had feveral child-ren, and had repeatedly mifcarried.

With a view to relieve the fymptoms above enumerated, a vein was opened in the right arm ; and other refources of art were had recourfe to. But on the fucceeding day a more violent attack of the difeafe took place. For the pulfe intermit-ted, though it ftill continued quick and hard ; un-lefs fhe lay with her neck erect, fhe could not breathe ; fhe expectorated a ferous livid-coloured fluid, and fhe paffed by ftool a bilious matter. On the third day the difficulty of breathing was ftill more confiderable : fhe became affected with fter-tor; her expectoration was in lefs quantity, and was very vifcid, and of a whitifh and fometimes yellowifh colour; fhe had frequent ftools; and her pulfe was ftill more irregular than it had been. Bleeding was again performed, and other remedies were employed : but in vain. For on
the

the fourth day all the fymptoms were much worfe : The pain of the cheft, which had been, as was obferved, dull and heavy, had now become acute, efpecially when fhe coughed or moved herfelf; and befides, a pain in the lower part of the belly, which fuggefted the idèa of labour having begun, fupervened. On the following night, amidft frequent irregularities and intermiffions, the pulfe began to flutter, the pains ceafed, the conftant reftlefsnefs of the whole body abated, the expectoration was fuppreffed, her ftrength was exhaufted; and thus, on the beginning of the fifth day, fhe died.

Appearances on Diffection.

ABDOMEN. The belly was opened immediately after death. The uterus was fomewhat inflamed ; and contained a female child, which was already dead. The vifcera of the child, as well as the abdominal vifcera of the woman, were in a natural ftate.

THORAX. The right lobe of the lungs appeared exceedingly inflamed and indurated, and had in its under portion an incipient abfcefs.

CASE XVIII. (XXI. 27.) .

A MAID fervant, aged twenty four years, of a good habit of body, and plethoric, who had been every winter fubject to a violent cough, having

been

been employed in much hard labour, became affected during the night with fever, ufhered in with
rigour followed by heat. Within twenty-fours,
pain in one part of the breaft, together with difficulty of breathing, fupervened. Along with thefe
fymptoms, fhe had conftant tickling cough, and her
pulfe was rather hard, and refifted the preffure of
the fingers almoft to the very laft. In the progrefs
of the difeafe the pain fhifted to the oppofite fide
of the thorax. She felt a fenfation of weight
within the breaft, and could lie on neither fide.
Blood was drawn, which feparated into a greenifh
coloured ferum, and a craffumentum with a buffy
coat on its furface; the part below this was very
black, and of a firm confiftence. Although blood-
letting was performed as foon as the pain came on,
and although that operation was repeated twice,
not only from the arms but alfo from the feet, on
the fame day, and other means of cure commonly
employed in fuch cafes were had recourfe to, fhe
died on the feventh day.

Appearances on Diffection.

THORAX. No fluid was effufed within the cheft,
nor did the lungs adhere anywhere to the pleura
except at the left fide, and there the connection
was neither ftrong nor extenfive. When the left
lobe of the lungs was feparated from thefe adhefions, for which purpofe it was preffed upon, a confiderable quantity of turbid ferum flowed out; but

whether

whether from the lobe itſelf, or from the interſtice between it and the pleura, within the limits of the connection, was uncertain, as neither the lungs nor the pleura exhibited the marks of any particular injury at that place; but the lobe was covered, even where there had been no adheſions, with a whitiſh and thickiſh membranous ſubſtance, ſuch as has been often deſcribed in the preceding caſes. A reddiſh coloured ſediment, ſimilar to what ſubſides from water in which freſh meat has been waſhed, adhered to the ſurface of the correſponding pleura.   In one part of the ſurface of the lungs, where there had been no adheſion, a kind of tubercle projected, which, when cut into, diſcharged a whitiſh ſerum like pus.   The lungs were not only heavy, but were alſo indurated in ſeveral places. When cut into, their ſubſtance, to a great depth, and to a conſiderable extent, was found of a denſe compact ſtructure like that of the liver.   In other places, they were almoſt of a red colour, and contained a large quantity of the whitiſh ſerous fluid, ſeen in the tubercle.   From theſe appearances, it was evident that death had been occaſioned by the inflammation and ſuppuration of the lungs. In the pericardium ſcarcely any fluid was obſerved.   No polypous concretions appeared in the heart.   The left ventricle contained ſcarcely any thing, but in the right there was a ſmall quantity of black coagulated blood.

C A S E

Case XIX. (xxi. 29.)

A MAID fervant, aged nineteen years, of a plethoric full habit, from expofure to cold during men-
ftruation, which for three months had been more
fparing than ufual, was affected with acute pricking pain in the breaft, attended with difficult refpiration. The pain was fixed under the left
breaft, and was aggravated when touched, fo that
fhe could not lie upon that fide. After having
been bled from the arm of the affected fide, fhe was
brought into the hofpital of Bologna. At that
time the fymptoms juft enumerated continued.
She lay upon her back. Her pulfe was quick and
fmall, and when preffed upon by the finger gave
little refiftance ; it was not however irregular. A
found like that of boiling water feemed to iffue
from the afpera arteria. She had a conftant hollow
tickling cough; her tongue was parched; and her
belly was moderately open. Along with thefe
fymptoms, fhe had at times flight delirium. Blood
was drawn from the right arm. When examined,
after having ftood for fix hours, it was found to
confift of a very little clear ferum of a golden colour, and of a yellow cruft two inches in thicknefs,
refembling rancid fat formed into a circle, the
edges of which were in contact with the fides of
the veffel. When the veffel was inclined to a fide
this

this cruft feparated into feveral laminæ, and the lower part of the blood formed grumous fubftances not unlike ftewed meat. After the venefection, on the fame day, a fmall quantity of blood, or rather a little ferum flightly tinged with blood, was difcharged from the uterus. All the fymptoms having become aggravated, the pulfe appearing deeper, as if the artery had retracted inwards, and a frothy fluid refembling water in which frefh meat had been wafhed having flowed out from the mouth, fhe died about the beginning of the feventh day of the difeafe.

Appearances on Diffection.

ABDOMEN. When the belly was opened, a ftrong fmell like that proceeding from inflamed vifcera, in a ftate approaching to gangrene was perceived. The liver was of a whitifh colour. The fpleen was tinged with a gangrenous blacknefs to a confiderable extent, but very fuperficially. The part thus tinged was the inferior portion only, fo that it did not touch the diaphragm which was perfectly found. A very bad fmell arofe from the uterus, its appendages, and the contiguous parts towards the back. A fmall quantity of reddifh coloured ferum furrounded thefe parts, and they were all externally livid from previous inflammation, which had extended to the lowermoft part of the rectum, and of the pudendum, as was evident from the rednefs of thefe

parts,

parts, the most minute vessels throughout them being as distinct as if distended with red wax. The inside of the thighs appeared variegated in an ugly manner, from the woman having been accustomed to put a small stove under her cloaths during Winter, in order to keep herself warm. The ovaria were not smaller than the ordinary sized testes of a man. They were soft, and contained a great number of vesicles of different sizes, filled as usual with a fluid, and supplied with ramifications of blood-vessels, which were very numerous in these ovaria. In each of them, there were also two small cells; one of these contained a minute black body, of a perfectly spherical figure, like a dried clot of blood. But in the other, there was nothing, as it consisted only of the membrane forming it, which was drawn together, and was of a green, and whitish yellow colour. One ovarium, at least, had the appearance of a chink next its surface, or at any rate, certainly exhibited marks of former injury. Two clots of blood were found in the vagina about the os tincae, and blood could be readily forced out from the fundus and cervix uteri. The internal surface of the fundus looked at first sight, like the mucous substance with which it had been covered. The situation of the Fallopian tubes, the appearance of the rugæ of the vagina, and of the carunculæ myrtiformes within the hymen, were perfectly natural.

I THORAX.

Thorax. The lungs were everywhere connected to the parietes of the thorax, and even to
the feptum tranfverfum, in fome places by means
of membranous fubftances; but in moft places,
and efpecially at the back and fides, they adhered by themfelves, their own membrane which
was found being united to the pleura. The pleura was only fomewhat thicker than ufual, and
eafily divifible into two laminæ. The fuperior
part of the left lobe was completely indurated,
refembled the compact fubftance of the liver, and
was of a whitifh colour. The fubftance of all the
other parts of both lobes, although entirely diftended with a frothy fluid, was in a natural ftate.
That fluid flowed out when thofe parts were cut
into; and at the fame time, a purulent-like matter was difcharged from many of the branches of
the bronchia; and from one of thefe, a white fubftance not in the leaft fluid, refembling a polypous
concretion, although perhaps it might have been a
fanguiferous veffel, was fqueezed out. In the trunk
of the afpera arteria, and in the larger divifions of
that canal, a whitifh cineritious matter, formed
here and there into concretions, was collected.
The tongue was covered with the fame kind of
matter; a whitifh and fomewhat thick fluid was
preffed out from the glands lying on the pofterior
part of the branches of the trachea; and all the
bronchial glands did not feem found. Between

the lobes of the lungs, in fome places, thofe membranous-like fubftances, fo often mentioned in the
preceding cafes, appeared. They were in this
fubject fomewhat thick and hard; but were not,
however, true membranes. In the pericardium
there was a quantity of reddifh-coloured watery
fluid. The ventricles of the heart contained no
blood; but fome polypous concretions, folid for
the moft part, and confifting of flefhy membranes,
as it were, and alfo in fome degree of mucus. Of
thofe which were in the right ventricle, one extended into the adjoining auricle and into the vena cava, and the other, which was round, went
into the pulmonary artery and its branches. One
only (alfo round) paffed through the left ventricle; from whence it ftretched on one fide into the
aorta, and on the other into the left auricle and
pulmonary vein. The cellular membranes, which
furround the trunks of the large veffels as they go
out of the pericardium, feemed to be of a mucous
confiftence, in confequence of watery fluid being
collected within them.

HEAD. The fauces and the neighbouring parts
had begun to fmell, as if from inflammation.
When the cranium was opened, the fame acid
kind of fmell which proceeds from the mouth of a
child affected with worms was perceived. White
polypous concretions were feen in the great falci-
form finus. The veffels of the pia mater were
somewhat

fomewhat more turgid than ufual. Between that membrane and the brain, a fmall quantity of watery fluid was obferved. The lateral ventricles contained a fluid of the fame nature, and of a reddifh colour. The choroid plexufes were pale, and had a few large hydatids adhering to them. The brain was fomewhat fofter than ufual, and the cerebellum much more fo.

C A S E XX. (LXIV. 2.)

A MIDDLE-AGED woman, affected with peripneumony, died in the hofpital of Bologna.

Appearances on Diffection.

ABDOMEN. A fubftance, of a roundifh form, of an inch in diameter, of a reddifh colour, invefted by a coat peculiar to itfelf, appeared in the adipofe membrane furrounding the left kidney. It was neither the glandula fuprarenalis, which at that fide was double, nor a fupernumerary kidney of a fmall fize, nor a lymphatic gland; but feemed, from its appearance on diffection, to be an additional fmall fpleen. For, when cut into, its circumference appeared of a bright red colour, as is often obferved in the fpleen; in other parts it was of a red brown colour: and although its fubftance refifted the knife fomewhat more than the fpleen does, yet every perfon who faw it, at once acknowledged it to be of the fame nature. The

left

left kidney was longer than the right; which was
not wonderful, fince it was furnifhed with a double pelvis, one lying above the other, each being
quite diftinct. The ureters, one of which proceeded from each pelvis, opened into the bladder
at the ufual part by an orifice proper to each, the
one being feparated from the other by a fmall
fpace. The internal furface of the fundus uteri
was red; and although veffels were diftinctly feen
in the membrane lining it, no blood was difcharged when the fubftance of the uterus was preffed
upon. In the internal furface of the aorta, fmall
white fpots were obferved, like incipient offifications. Within the iliac arteries, parallel lines, in
a longitudinal direction, appeared on the internal
furface, which could not be obliterated even by
drawing out the fides of the veffel.

THORAX. A great part of one of the lobes of
the lungs was fwelled, indurated, heavy, and internally of a firm compact fubftance; which was
not only of a pale red colour, like the liver when
boiled, fuch as is often feen in inflammation of
the lungs, but was alfo of a whitifh colour, apparently from its containing purulent matter, though
it was probably frozen, as intenfe froft prevailed
at the time the body was opened. The cellular
membrane of the afpera arteria, on the pofterior
part, was fo much diftended with pent-up fluid,
that the glands commonly feen there could not
be

be diftinguifhed. The blood returning from the belly paffed through the diaphragm by two orifices, contiguous to each other, inftead of one. In the heart, the orifice of the coronary vein was not furnifhed with a membranous valve ; but was clofed on the right and left fide with numerous flender parallel filaments, reaching from the upper to the lower part, in fuch a manner, that the blood could pafs between filament and filament, but more readily through the middle of the orifice, where the filaments were almoft wanting.

Case XXI. (xx. 61.)

A woman above fixty years of age, who was fat and plethoric, was feized with a pain fituated chiefly in the right fide of the thorax ; together with violent fever, great thirft, forenefs and laffitude over the whole body, and painful breathing. Venefection was performed, and other means of art employed ; but in vain : For the laffitude increafed, fhe became infenfible, and had low delirium. At length her expectoration, which had hitherto been in moderate quantity and not vifcid, became copious, and had a purulent appearance *. Her refpiration grew more laborious. Delirium took place ; and fhe died on the eighteenth day of the difeafe.

Appearances

---

* In the original the expreffion is, " Tandem fputum, quod antea modicum erat, nec glutinofum, rotundum ejicitur et purulentum."

Appearances on Diffection.

THORAX. The right lobe of the lungs adhered
to the pleura, and contained within its fubftance
an abfcefs, round which there was great inflammation, and from which a quantity of purulent
matter was difcharged when the lobe was feparated from the pleura. That membrane itfelf, and
the left lobe of the lungs, were found. In the
right ventricle of the heart there was a fmall polypous concretion; and in the other a fubftance
of the fame kind, in an incipient ftate.

CASE XXII. (xx. 35.)

A YOUNG man, nearly twenty-four years of
age, was feized with a pricking pain in the left
fide of the thorax; attended with difficulty of
breathing, great thirft, violent cough, with no expectoration. ' He lay on the affected fide, and felt
lefs uneafy when his head was low. At length,
the difficulty of breathing having every day become greater, he died on the fixteenth day of the
difeafe.

Appearances on Diffection.

THORAX. The left lobe of the lungs was everywhere ftrongly connected to the neighbouring
parts: it was inflamed, and, towards the fcapula,
in a ftate of fuppuration. The pericardium was
fo much diftended with fluid, that it might very

juftly

juftly be deemed dropfical. Its coats were very much thickened; and certain white concreted fubftances adhered in fome places to its internal furface, as well as to the outfide of the heart. The ventricles contained fluid blood, and alfo two polypous concretions; that in the right ventricle being larger than that in the left.

CASE XXIII. (xx. 49.)

A WOMAN, aged fixty years, of a flender make and fanguineous temperament, was feized with a violent fever, dry cough, and an acute pricking pain in the left fide, of fuch a nature, that when fhe lay on that fide it was much alleviated. After blood-letting the pain remitted fo much, that fhe could lie eafily on any fide. The fever, neverthelefs, and thirft, continued. She fometimes expectorated a thick matter. The difficulty of breathing became aggravated; and at length, on the thirteenth day of the difeafe, fhe died.

Appearances on Diffection.

THORAX. The right lobe of the lungs was quite unconnected with the pleura. It was of a whitifh colour, marked with black fpots, of a firmer confiftence than flefh, and contained fome tubercles filled with fanies. The pleura was quite found. The left lobe of the lungs adhered to the pleura,

pleura at the fuperior part, but was much lefs difeafed than the right; for it was only covered with
black fpots, and towards the throat, where it was
pretty hard, contained a fmall abfcefs. In the
pericardium there was an ounce of ferous fluid.
A large polypous concretion was feen in the right
ventricle of the heart; the greateft part of which
filled the cavity of the correfponding auricle, from
whence it extended into the contiguous veffels.
In the left ventricle there was another polypus,
of a fmaller fize.

C a s e XXIV. (xxi. 23.)

A man, aged about forty years, was feized with
an acute pricking pain in the right fide, attended
with fever and intenfe heat of the body. He was
brought into the hofpital of Bologna, on the beginning of the fourth day of the difeafe. He had
then a fmall, quick, and fomewhat irregular pulfe,
and alfo frequent and weak refpiration. He lay
upon his back, appeared drowfy, and his intellects
were deranged. He had been bled previous to
his admiffion into the hofpital. He had paffed a
reftlefs night, and his refpiration had become more
difficult. On the morning of his admiffion he
breathed with ftill greater difficulty; and ftertor
having fupervened, he died about the end of the
fame day.

1   *Appearances*

Appearances on Diffection.

ABDOMEN. On the fmall inteftines, in a very few places, a flight incipient inflammation was obferved. The liver appeared fomewhat indurated. It was livid, both on its edge, and alfo to a confiderable extent on its concave furface. In the latter part the livid hue was fuperficial, but not in the former. The fpleen was fo flabby, that it was very readily broken down by the fingers.

THORAX. From the pleura, near the left fide of the fpine, three or four tubercles, of a white colour, and of the hardnefs of bone, projected. In other refpects, that membrane was found. The left lobe of the lungs, although it appeared uninjured, difcharged, wherever cut into, a yellow coloured fluid. The right lobe was greatly enlarged, very heavy, and much indurated; and refembled the fubftance of the liver, except in fome parts, to a confiderable extent, where it was of a white colour, and in a ftate approaching to putrefaction. From it, too, a fluid fimilar to that in the left lobe, and in greater quantity, flowed out. In all the orifices of the heart, polypous concretions were obferved. Of thefe, the fmalleft was fituated in the left auricle, and the largeft in the right; and both were accompanied with grumous blood. The polypi, in the pulmonary artery, and in the aorta, were of an intermediate fize between thefe two.

VOL. I.　　　　M m　　　　HEAD.

HEAD. The veffels of the pia mater were fome-what turgid. Under this membrane, a watery fluid was feen; and the ventricles contained a fmall quantity of reddifh coloured fluid of the fame kind. The plexus choroides were neither pale, nor free from hydatids.

CASE XXV. (xx. 56.)

A MAN, about fifty years of age, who had for many years laboured under a leprofy of the right thigh, was affected with fore throat. After this had continued for two days, it terminated in a particular pain in the back. To this a dry cough, great thirft, and difficulty of lying on the right fide, though he could eafily lie on the left, fuper-vened; and, befides, he felt a certain oppreffive pain, furrounding like a belt the lower part of the thorax. Although, during the latter days of his life, the febrile fymptoms appeared mitigated; and though no other marks of inflammation of the cheft than thofe already enumerated occurred, he died on the ninth day from the time he had been confined to bed.

Appearances on Diffeftion.

THORAX. The left cavity of the cheft was filled with purulent ferous fluid. A membranous fubftance, which was a concretion of this fluid, adhered to the pleura of that fide, in fuch a manner

ner as to refemble a difeafe of that membrane.
The pleura, which lay under this concreted fubftance, was inflamed ; as was alfo the cafe, though
in a flight degree, with the lungs. Some very
fmall polypous concretions lay hid in the large
veffels about the heart.

C a s e XXVI. (xxi. 24.)

A MIDDLE-AGED woman, who had mifcarried a-
bout the third month of pregnancy, although fhe had
loft a moderate quantity of blood from the uterus, and
had alfo had a vein opened, was on the eighth or
tenth day after mifcarriage affected, without any
apparent caufe, with internal inflammation of the
cheft. On this account fhe was brought into the hofpital at Bologna. She then complained of pain in
the cheft, and of difficulty of breathing, attended
with fever. She lay always on her right fide, as fhe
could neither lie on her left nor on her back. The
pain was internal, fo that it fhould not have been
increafed by touching the affected part ; fhe could
not, however, defcribe accurately its feat. She had
cough ; but expectorated nothing, at leaft nothing
which had any reference to the difeafe. Deafnefs
and pain in the ears fupervened. As fhe had been
formerly bled from the arm, a vein was now opened in the foot : but thefe and other means proved

ineffectual;

ineffectual; for fhe died about the thirtieth day after mifcarriage.

Appearances on Diffection.

ABDOMEN. The belly was fwelled, from the e-
pigaftric region downwards; and when the fwelling was preffed upon, air was forced out at the
mouth. The liver was of a prodigious magnitude,
and had forced the ftomach, which was diftended
with air, down into the umbilical region. It was
throughout very much indurated, and of a colour
which, although it approached to whitenefs, was
not very different from the natural one. The bile
in the gall-bladder was almoft black. The fpleen
was large, but nothing in proportion to the fize of
the liver. The parietes of the uterus were fomewhat thicker than ufual, but contained no blood :
for, though cut into, and preffed upon by the fingers, not a drop appeared. The internal furface
of the fundus was livid. One of the Fallopian
tubes had hydatids hanging from its fimbriæ, by
which its orifice at that extremity feemed clofed
up. Each tube contained fuch a quantity of the
white puriform fluid which lubricates their external furface, that, when preffed upon at the extremity next the uterus, the fluid was forced out, and
the internal orifice was thereby rendered diftinctly
vifible. Within the middle of one of the lobes, a
fmall body, of a black colour, and of the fhape of
a grape, was feen. This had a minute ftalk attached

tached to it, which feemed to be, and probably
was a fmall clot of blood. The ovaria were unequal in their furface, and each was marked with a
black fpot. Under thefe fpots, a minute cyſt,
filled with a black globule, lay. In the centre of
one of thefe globules, there was another hollow
globule, of a fmaller fize, and of a mixed black
and yellow colour. The remaining parts of the
uterus and its appendages had a gangrenous fmell.
A fmall quantity of reddiſh coloured turbid watery
fluid was found within the pelvis.

THORAX. Both fides of the cheft contained a
little yellowiſh turbid watery fluid, of which there
was more in the left than in the right fide. The
lungs adhered almoft every where to the pleura.
When feparated, a whitiſh pellicle, eafily lacerated followed; this was certainly not the membrane
of the lungs, for that lay under it and was quite
found; but whether it belonged to the pleura, as
it appeared, or was of the fame nature with thofe
membranous fubftances feen in fimilar cafes interpofed between the pleura and the lungs, could
not be accurately determined. The lungs were
inflamed, efpecially at the pofterior part, and their
fubftance was thickened, was fomewhat indurated, and was in fome places of a blackiſh colour.
In the pericardium there was a fmall quantity of
a reddiſh turbid watery fluid. In each ventricle
of the heart, polypous concretions of a mucus-
like

like confiftence, and of a whitifh yellow colour were feen.

HEAD. A fmall mucus-like concretion, like that noticed in the ventricles of the heart, was found in the fuperior finus of the falx. Under the pia mater there was a little watery fluid, but there was fcarcely any in the ventricles. The brain was not at all flabby; and the choroid plexufes were in a natural ftate. The pineal gland was of fuch a fize as to equal nearly that of an ordinary grape; and being flightly touched with the knife, difcharged a turbid watery fluid, together with a very fmall quantity of a yellowifh mucous matter, after which it decreafed in fize. When the ears were examined, the membrana tympani of each was found to be of a black colour, and exceedingly flaccid; and alfo the maftoid cells adjoining to the tympanum were more moift than ufual. In one of the tympana, there was a kind of purulent matter; and in the contiguous part of the occiput on the outfide, efpecially at the left fide, all the cells in the integuments were filled with a watery mucous fluid. On each fide, the parotid gland and meatus auditorius were found.

CASE XXVII. (xx. 13.)

A BUTCHER, aged fifty years, became affected
with

with a pain in the right fide of the thorax, extending to the middle of the fternum. He lay on his back, coughed much, and expectorated little. He was obliged to hold his neck erect that he might breathe. On the feventh day he died.

Appearances on Diffection.

THORAX. The right lobe of the lungs adhered fomewhat to the fternum, and very clofely to the mediaftinum : its fuperior part was entirely indurated. The left lobe was on its pofterior part of a black colour. On the left fide alfo, the thoracic cavity contained a fmall quantity of ferous fluid. The pericardium was filled with the fame. In the right ventricle of the heart a polypous concretion was feen.

The blood in this body approached nearer to fluidity than to coagulation.

CASE XXVIII. (xx. 32.)

AN unmarried woman, aged fixteen years, of a cacheclic habit of body, whofe menfes had been fuppreffed for eight months, who had a fpitting of catarrhous matter, felt difficulty in breathing when walking, and was accuftomed to complain of a certain fenfation of heat, and pain in the left fide of the thorax; was fuddenly affected with fuch extreme difficulty of breathing, that fhe could not poffibly lie in bed. This was attended

with

with the expectoration of matter tinged with blood, with the fenfation of an oppreffive weight in the feat of the pain, with a hard pulfe, and coldnefs of the extremities. At length, on the fourth day from the beginning of the laborious refpiration, fhe expired.

Appearances on Diffection.

ABDOMEN. The gall-bladder was very fmall, its coats were much thickened, and it contained little or no remains of bile.

THORAX. A quantity of limpid ferous fluid was found in both cavities of the cheft. This fluid, when fet at reft, exhibited the fame appearances as blood does in a fimilar fituation; for in its middle a fubftance like jelly appeared, feparated from, and furrounded by, the reft of the fluid. When expofed to heat, it coagulated in the fame manner as the ferum of the blood does. The whole left lobe of the lungs was exceedingly indurated. The right ventricle of the heart contained a very large polypous concretion. In the left, there was alfo a fimilar body, but of a fmaller fize.

CASE XXIX. (xx. 33.)

AN old man, aged feventy-four years, became affected with a pain of the right fide, which was very diftreffing, rather from a fenfe of weight, than

r   from

from any other caufe. It was more uneafy if he lay upon the oppofite fide, and therefore he lay on the right fide. Along with this pain he had troublefome cough, accompanied with bloody expectoration, pain in his head, watching, and more efpecially violent fever, attended with hard, frequent, but not very full pulfe. He died on the third day.

Appearances on Diſſection.

THORAX. The right fide of the cheft was completely filled with ferous fluid. The lobe of the lungs on that fide, was of a black colour, and was inflamed, indurated, and of a very compact fubftance. It was eafily feparable from its external-membrane, and towards the fcapula was connected to the pleura by fhort membranous bands. The left lobe was fcarcely injured. In the right ventricle of the heart, a moderate fized polypous concretion, belonging principally to the correfponding auricle, was obferved. The left contained a fmaller one. The former of thefe concretions was continued into the vena cava and pulmonary artery; and the latter extended for a fhort way into the pulmonary vein, and for a confiderable length into the aorta. Both concretions were accompanied with coagulated blood.

C a s e XXX. (xx. 39.)

A man, of about fixty-fix years of age, who had long expectorated catarrhous matter, having been expofed to cold, was feized with an acute pricking pain in the left fide, attended with confiderable cough, and with fever. From the very beginning of the pain, he expectorated a large quantity of thick yellow matter, marked with fmall ftreaks of blood. At length the expectoration having been fuppreffed, he died on the feventh day.

Appearances on Diffection.

Abdomen. The fpleen was of a reddifh colour.

Thorax. A quantity of fluid, like cow-milk whey, was found in the left cavity of the thorax. The correfponding lobe of the lungs was very much indurated, and adhered ftrongly to the mediaftinum, and to the pleura invefting the ribs. The right lobe was connected in the fame manner to the mediaftinum and to the pleura lining the diaphragm, and more efpecially to that under the upper ribs at the fore part. A cancerous ulcer at that place lay concealed within the lobe : this was probably the feat of an old difeafe. In the left ventricle of the heart, there was a fmall polypous concretion. The right ventricle contained one, which was much larger, adhering to a quantity of coagulated blood.

3   C a s e

Case XXXI. (xx. 30.)

A CLERGYMAN, who had just entered his twenty-third year, had about three years before been affected with acute fever, attended with a running from the parotid glands. After having recovered from this indisposition, he fell into a double tertian fever, which continued for a long time. When this at length ceased, he remained in some measure emaciated, with a pale countenance; was troubled with difficult respiration, and occasionally disturbed sleep; and his urine was almost always of a red colour. To these symptoms, an acute fever at last supervened; and, on the second day after, a pain in the right side, below the false ribs, and below the ensiform cartilage, took place. This pain was encreased on touching the affected part. On the first day of the disease, vomiting and diarrhœa attended; and he was also affected with cough, which was at first accompanied with some expectoration, but in a very short time became quite dry. As he could not lie on either side, he lay on his back. He complained of the sensation of intense heat towards the right kidney. His pulse was weak, quick, sharp, irregular, and intermitting. The disease having continued to encrease in violence, the difficulty of breathing having become every day

more

more confiderable, and his pulfe having grown very weak; he died about the end of the feventh day.

Appearances on Diffection.

ABDOMEN. All the vifcera were perfectly found, and in a natural ftate; except the fpleen, which was four times larger than ufual.

THORAX. The left cavity of the cheft contained more than two pounds of liquid ferum. The right cavity was filled with a thicker fluid; fome part of which had become concreted in fuch a manner, as to form the appearance of membranous bodies floating through it. The lungs were unconnected with the pleura. The right lobe, although not much encreafed in fize, was entirely indurated from previous inflammation. The pericardium contained a great quantity of ferous fluid, which rendered it much larger than ufual. The right ventricle of the heart was filled with coagulated blood, together with a fmall polypous concretion. The right auricle, alfo, was much diftended with coagulated blood; and the left ventricle contained a fmall quantity of the fame kind of blood.

CASE XXXII. (xx. 36.)

A YOUNG man, about twenty-two years of age, after having complained of pain in his belly, which

had

had ceafed, was affected with pain in the thorax, attended with difficult refpiration, cough, and troublefome thirft. He lay conftantly on his face, but in fuch a manner that he inclined towards the right fide. He held alfo his head low, as thofe who have the pericardium diftended with fluid generally do. All thefe fymptoms having become aggravated, he died on the fixteenth day.

Appearances on Diffection.

THORAX. The left fide of the cheft contained a quantity of limpid fluid. The correfponding lobe of the lungs was perfectly found, and everywhere unconnected. The right lobe was indurated, and ftrongly connected with the neighbouring parts, and efpecially towards the fcapula, where a fluid like cow-milk whey, containing fome concretions refembling the white of a boiled egg, was obferved. Within the pericardium, which was much diftended, and occupied a large portion of the thorax, a fluid of the fame kind, with fimilar concretions, was found. Thefe concretions adhered to the internal furface of the pericardium, and to the outfide of the heart. Two polypous fubftances appeared in the ventricles of the heart ; that in the right was larger than that in the left.

CASE XXXIII. (xx. 59.)

A WOMAN, aged fixty-four years, was feized with

with a pain in the right fide of the thorax, in fuch
a manner that fhe lay with difficulty on that fide.
and could not bear to have the affected part touched. Her refpiration was frequent. She had a
dry cough, and a quick, fmall, feeble pulfe. On
the feventh day, a fweat broke out about the head.
On the ninth day, her ftrength having become
quite exhaufted, fhe died.

Appearances on Diffection.

THORAX. The right lobe of the lungs appeared
to be fo fwelled, that it filled the whole cavity on
that fide. It adhered flightly to the pleura, by
means of a kind of membranous fubftance, which
was interpofed. This, however, was in fact, nothing but a concretion from ferum, fpread out in
fuch a manner as to refemble a membrane, and to
make the pleura appear corrupted *. The pleura in this cafe was found. The fame lobe at its pofterior part, was indurated and inflamed. The
left lobe was connected in feveral places to the
pleura ; but was in other refpects uninjured. In
the pericardium there was a great deal of watery
fluid. Within the heart feveral polypous concretions were found. In the right auricle, a pretty
large one, extending into the vena cava, and in the
correfponding ventricle a fmaller one, continued into

* An appearance of this kind feems to have impofed upon Ri-
verius, when he faid he faw the pleura corrupted in a cafe of pleurify.

to the pulmonary artery, were feen.   In the left
ventricle, there were alfo two polypi of different
fizes, the larger of which ftretched into the aorta,
and the fmaller into the left auricle.

Case XXXIV.   (xx. 53.)

A man-servant, aged fifty-five years, complained of pain in the middle part of his cheft.   He
could lie on his back, and alfo in fome meafure on
his left fide.   He was obliged to have his neck e-
rect that he might breathe.   He died on the fixth
day of the difeafe.

Appearances on Diffection.

Thorax.   The left lobe of the lungs, on the pofterior part, had become very much indurated, and
adhered ftrongly to the pleura, even where that
membrane invefted the diaphragm.   The pleura
was fomewhat reddened.   The right lobe was
found; although a fmall quantity of ferous fluid,
like pus, was obferved in that fide of the thorax.
The pericardium contained fome turbid fluid.   In
the right ventricle of the heart, together with a
polypous concretion, coagulated blood was found.
In other parts of the body the blood was in fome
degree fluid.

Case

Case XXXV. (xx. 47.)

A young man, about twenty-fix years of age, was feized with an acute pricking pain in the right fide of the thorax, attended with fever and cough, with little expectoration. About the eighth day, delirium, particularly violent during the night, fupervened. The pain ceafed; but the difficulty of breathing became aggravated. He could lie eafily on either fide. On the tenth day he died.

Appearances on Diffection.

THORAX. Both lobes of the lungs were much indurated, and were towards the back connected to the pleura; to which alfo the left lobe adhered laterally. A white membranous fubftance, like a foft flaccid reticulated body, was fpread over the whole of the pleura and lungs, which in fome parts were thereby connected together. In both fides of the cheft, but more efpecially in the right, a large quantity of fluid, like cow-milk whey, was found. This fluid did not feparate into parts, when fet at reft; but, when expofed to heat, coagulated, like the ferum of the blood. The pericardium was diftended with ferous fluid. A polypous concretion was feen in each ventricle of the heart; that in the left was the fmalleft.

Case XXXVI. (xlv. 16.)

A MIDDLE-AGED woman, of a pretty good habit of body, and of a moderate stature, having been previously affected with catarrh, was seized with fever; on which account she was brought into the hospital of Bologna. Along with the fever, which was very violent, she had great difficulty of breathing, flushed face, a most distressing sensation of weight in the thorax, and her pulse was somewhat hard. She was exceedingly anxious to expectorate, but she could not do so. Her pulse became low and intermitting; and her respiration so difficult, that at last she could not lie. She neither complained of uneasiness nor acute pain in the back; nor was she ever affected with palpitations of the heart, nor delirium. Every necessary means were employed, but in vain; for she died on the fifth day from the beginning of her feverish complaint.

Appearances on Dissection.

ABDOMEN. The spleen was large. The liver was so very large, that, filling up the left hypochondrium, as well as the right, it had depressed the stomach; consequently a part of the œsophaphagus terminating in that organ, appeared two fingers breadth below the diaphragm. But, except their size, nothing uncommon appeared, ei-

ther in the fpleen or liver. A thick oblong polypous concretion was found within the inferior vena cava. The uterus lay forwards, and was fomewhat nearer the left than the right fide. The
ovaria were very long and flender, of a white colour, and indurated; and were connected to the
uterus by thicker ligaments than ufual. The
veffels which run through the broad ligaments
into the uterus, were very much diftended with
black blood, and were here and there varicofe.
When an incifion was made from the upper part
of the uterus to the lower part of the vagina, the
fundus and cervix uteri were found full of mucus;
which was almoft tranfparent like jelly, tinged
with no colour, and thinner than that ufually feen
about the os uteri, which in this fubject was not
wanting. When the mucus at the fuperior part
was removed, a very minute excrefcence, almoft
of a circular form, and of a reddifh brown colour,
was obferved to project from the internal furface
of the fundus; and the thick mucus at the lower
part of the uterus having been removed, the inferior part of the cervix appeared of an unequal furface, from unufual fhort lines, of a red colour,
which were placed longitudinally, and projected
fomewhat. The vagina, although it was not deftitute of rugæ, from the middle downwards, was,
in proportion to the fize of the woman, who was
as already mentioned of the middle ftature, fomewhat

what longer and broader than it ought to be; and contained a wooden ring peſſary, which was a mark of former prolapſus. This peſſary was of an oval form; and was placed in ſuch a manner, that its longeſt axis correſponded with the length of the vagina, and the ſhorteſt with the breadth; one end of the oval being turned towards the os tincæ, and the other towards the orificium vaginæ. The ſhort diameter was ſo long as to diſtend both ſides of the vagina; which, at the part preſſed upon by the inſtrument, projected into an excreſcence, of the form and ſize of a large almond kernel, of a cartilaginous hardneſs, and of a white colour, except that one of them was livid in the middle. · Theſe excreſcences ſeemed to threaten an approaching change, from a ſcirrhous nature into ſomething worſe.

THORAX. The lungs, although turgid, adhered ſtrongly almoſt everywhere to the pleura lining the ribs, eſpecially on the left ſide. From the ſame ſide of the thorax, a ſerous fluid, which would have appeared to be white from an admixture of pus, had there been any veſtige of purulent matter at that part, flowed out in great abundance: for it had been collected in ſuch a quantity, that the diaphragm on that ſide, when looked at from the abdomen, appeared convex inſtead of concave. Part of the ſame kind of matter was included between the left lobe of the lungs and the

pleura

pleura invefting the ribs, nearly about the middle
of the dorfal vertebræ, to a great extent.   Within that fpace white concretions, like thick membranes, appeared, adhering both to the lungs and
pleura.   The lobe was, at that part only, of a
harder and more compact fubftance than natural.
A great part of the pleura on both fides was of a
rofy colour.   The pericardium was large, and filled with the fame kind of fluid found in the cavity of the cheft : fo that at firft fight it might
have been miftaken for a large open abfcefs,
inftead of the pericardium.   When the fluid was
difcharged, the internal furface of the pericardium, and the external furface of the heart, auricles, and large veffels, appeared of a pale livid
colour, and covered over with a certain white cineritious matter, like plaifter newly laid on a wall ;
but which, in fact, was found to confift of polypous concretions, forming a thick flaccid membrane, that could be eafily feparated and readily
lacerated.   When this preternatural membrane
was removed, the parts under it appeared in a natural ftate, and of a proper colour ; except that
the pericardium was thickened and fomewhat red,
as if from phlogofis, for it could not be called inflammation.   The heart, which was larger than
ufual, contained on both fides black blood, fuch as
had been feen in different parts of this fubject ;
and

and befides, fome round polypous concretions in the right ventricle and correfponding auricle.

Head. In the medullary fubftance of the brain, wherever it was cut into, and on the furface of the lateral ventricles, veffels fomewhat turgid with blood appeared. The lateral ventricles contained fome ferous fluid, of a dirty yellow colour.

Case XXXVII. (xxi. 36.)

A young man, aged twenty-five years, of a complexion rather pale, who had formerly been affected with ftrumous fwellings, and alfo, it was faid, with lues venerea, having overheated himfelf with running (for he was a footman), was feized with rigour and fever, attended with acute pricking pain, which he faid he felt almoft over the whole breaft properly fo called, but more efpecially at the lower parts of the cheft. His back alfo was pained ; and he complained of forenefs over the whole body, aggravated, he alleged, on being touched. He could lie only on his back. His pulfe was quick and fmall. He had no thirft; he felt great heat internally ; his refpiration was difficult ; and he expectorated a reddifh fluid matter. On the fourth day of the difeafe, he had fome bilious ftools. On the eighth day, after having paffed

a

a great quantity of urine, he became covered with a clammy fweat, and expired.

Appearances on Diſſection.

ABDOMEN.   The ſtomach and the inteſtines, eſpecially the colon, appeared much diſtended with air.   The lower part of the ſpleen was tinged with a particular blacknefs, to the depth of an inch and an half, juſt as if incipient gangrene had fupervened.   The liver was very large, and of a whitiſh colour.   The gall-bladder, along with a little bile of a whitiſh yellow colour, contained more than ſeventy calculous concretions.   The largeſt of theſe (of which there were very few) did not exceed the fize of a bean; and the fmalleſt were not lefs than a pepper corn.   The former were of an oval figure, and were flat on one ſide.   The latter were of the form of a cube.   Internally they feemed compoſed of a kind of minute grains moiſtened with bile.   Theſe grains were inveſted all round, by a cruſt, which was at leaſt double, and was moſtly of a greeniſh colour, but in ſome places white.   When expoſed to the fire, all of them preſerved the flame when they had once caught it, and now and then emitted ſparks with a noiſe.   The right kidney was exceedingly flabby.   The trunk of the aorta, both in the belly and in the thorax, as high up as the curvature, was fmall in proportion to the fize of the body and the

the other vifcera; all which were of a tolerable bulk.

THORAX. The right fide of the cheft contained a turbid fanious ferous fluid. Both lobes of the lungs adhered almoft everywhere to the contiguous parts, and even to the diaphragm. The connections of the left lobe, at the anterior furface, were formed by filaments only; the remaining part of the fame lobe, and the right one too, throughout their whole extent, were not only firmly connected to, but alfo even feemed to form the fame fubftance with the contiguous parts, by means of a thick membranous body interpofed between them and the pleura. This membranous fubftance was of a white colour, exceedingly tough; and in fome places, on the right fide, was half as thick as one's little finger. On the left fide, it was much lefs thick and white. On the lungs being forcibly feparated from their adhefions, this fubftance followed. When feparated from the lungs, which could be very readily done, this membrane appeared found and entire; infomuch fo, that, unlefs the pleura had been obferved uninjured in its natural fituation on the left fide, (through which the adjoining intercoftal mufcles appeared redder than ufual), it might have been miftaken for that membrane. On the right fide, however, another membrane fimilar to the former (which at that part had been torn off, along with

2   the

the lungs), except that it was lefs thick, appeared inftead of the pleura. When this membrane was drawn off, the intercoftal mufcles were feen under it, of a white colour; fo that, at this place, it was probable the pleura had become thickened in confequence of difeafe. The pleura, where it covered the fterno-coftalis mufcle, which was inflamed, was femi-putrid, and eafily lacerated by the flighteft touch of the finger. The flefhy part of the diaphragm, all round its middle, where the lungs adhered, feemed inflamed. The left lobe of the lungs exhibited in one place a white fubftance, compofed as it were of tartarifated grains. In almoft all the other parts, it was only indurated, compact, and heavy; and it was leaft fo on its anterior furface. The right lobe was much harder, more compact, and heavier than the left. In the pericardium there was a little more fluid than ufual; and it had a turbid appearance. A whitifh, yellow, rather flabby polypous concretion was feen in the right ventricle of the heart, extending through both orifices of that cavity. Similar concretions went out at each orifice of the left ventricle. Thofe which extended from this ventricle into the pulmonary vein, and from the right into the pulmonary artery, were divided into ramifications refembling divifions of thefe veffels.

CASE

to making .. 3 . 9
pasthood buried 4 .. 0 . 10
Ribbon sewing silk . 0 .. 7
5 2

CASE XXXVIII. (xxi. 34.)

A MAN, whofe trade was that of fifting wheat, an occupation very injurious to the lungs, after having been affected with pleurify, from which he was recovering, became again indifpofed. He complained of a violent pain of the thorax ; and was compelled, in order to breathe, to fit up in bed. He had an exceffive cough, attended with the expectoration of a fmall quantity of vifcid bloody matter. His face was flufhed; his pulfe was hard, exceedingly irregular and intermitting; and he was at times affected with convulfive tremors. He was bled two or three times; the blood drawn had a thick buffy cruft. His lower extremities became œdematous. Under thefe fymptoms he died, on the eleventh day.

Appearances on Diffection.

ABDOMEN. A confiderable quantity of turbid yellowifh watery fluid was found in the belly. The liver was indurated : externally it was. of a pale livid colour, and internally it was variegated by numerous red particles.

THORAX. Both fides of the cheft, and efpecially the right one, contained a great quantity of turbid yellow coloured fluid; amidft which there were concretions, fimilar to what are feen floating at the bottom of a cafk of wine. The pleura appeared at the fides, and efpecially at the left, which was probably the feat of the former pleurify, to be fomewhat redder than ufual. The lungs, which adhered nowhere to the pleura, had the lower part of the right lobe indurated and turgid; and at that part the fubftance appeared thick, and of a brownifh purple colour. In other parts they were of a foft confiftence. A bloody fluid was preffed from the bronchia on both fides into the trunk of the afpera arteria. The pericardium, before it was cut into, appeared much larger than ufual. When opened, this was found to be owing not fo much to a quantity of fluid, of the fame kind with that in the thorax, which it contained, as to the fize of the heart; the parietes of which indeed were not enlarged; but its ventricles (efpecially the right one) were greatly diftended, and filled with a large quantity of black blood. The blood certainly was not very fluid; but no polypous concretions appeared in the heart, except a fingle thin incruftation in the right ventricle.

CASE XXXIX. (xx. 7.)

A WOMAN, aged twenty-feven years, who though fhe had been married for four years, had never conceived, became affected with pain in the left fide of the thorax, together with difficulty of

I   breathing,

breathing, and violent cough, attended with fome expectoration. Under thefe fymptoms fhe died.

Appearances on Diffection.

ABDOMEN. The fluid of the veficles of the ovaria appeared coagulated, as if it had been boiled.

THORAX. In the left cavity of the cheft, a quantity of white coloured ferous fluid was found; and the pofterior part of the lungs in the fame cavity was inflamed. When cut into, although no abfcefs could be diftinguifhed, fanious matter mixed with blood flowed out, and black fpots were feen here and there throughout its fubftance.

CASE XL. (XX. 20.)

A PRIEST, nearly thirty years of age, was affected with difficult refpiration, and expectorated a large quantity of matter. At firft he complained of pain in the right, and then in the left fide of the thorax. He died on the tenth day.

Appearances on Diffection.

THORAX. Both cavities of the cheft contained ferous fluid, but not in confiderable quantity. A part of this fluid had become concreted; and, like a pale coloured coat, covered every where the furface of the lungs. The right lobe was exceedingly red at the pofterior part; it was alfo indu-

rated,

rare, but lefs fo than it generally is in cafes of peripneumony. About the middle of its internal fubftance, purulent matter had begun to be formed. The left lobe exhibited likewife marks of incipient inflammation on the back part. The pericardium was thicker than ufual, and was diftended with a yellow coloured ferous fluid. In confequence of the concretion of a part of this fluid, a reticulated kind of fubftance was extended, not only over the internal furface of the pericardium, but alfo over the external furface of the heart. When this fubftance was preffed upon, fmall drops of ferum flowed out. The right ventricle of the heart contained a polypous concretion.

CASE XLI. (xx. 2.)

A CARMAN, aged about fifty years, had been troubled with feverifh fits for the fpace of a year, and more lately alfo, had been affected with peripneumony, dry cough, difficult refpiration, and in fome degree, delirium. He could not, neverthelefs, be perfuaded to confine himfelf to bed, but applied to bufinefs as ufual. It was not, therefore, till about the fifth or fixth day from the firft attack of peripneumony, that he came to the hofpital of Bologna, to which he walked. After his admiffion, the above mentioned fymptoms became

came fo much aggravated, that he died within twenty-four hours.

Appearances on Diffection.

ABDOMEN. A fmall quantity of limpid watery ferum was found in the cavity of the belly. The fpleen appeared three times larger than ordinary.

THORAX. The right cavity of the thorax contained fome ounces of turbid ferous fluid. The fuperior lobe of the lungs, on that fide, was entirely inflamed, efpecially towards the pofterior part. When cut into, very fmall abfceffes, containing fanious matter, were found everywhere difperfed here and there throughout its fubftance. The pleura was found. The cavity of the pericardium was half filled with its own proper fluid. Polypous concretions were found in the heart. Thofe in the auricles were the larger : thofe in the ventricles, at the orifices of the arteries, were the fmaller ; and of them the largeft was fituated in the right ventricle.

CASE XLII. (xxi. 17.)

AN old man, above fixty years of age, became affected with fever, and at the fame time with an acute pricking pain in the anterior part of the right fide. He lay upon his back. His tongue was parched, and his pulfe full and frequent. After having been treated for fome days in the ordinary manner,

manner, in the hofpital of Bologna, he feemed to
be fo much relieved of fever and pain, that the fe-
nior phyfician allowed him to indulge freely in
the common diet of the houfe; and three days after ordered him a purge, according to cuftom.
Having, in confequence of this permiffion, eaten a
great deal, the fever and pain of the fame part returned on the fucceeding night. His pulfe became
hard, frequent, full, and vibrating; and continued fo till within a few hours of his death. His
refpiration was hurried; but was not very bad.
Having fpontaneoufly raifed himfelf into the fitting
pofture, although he appeared as if he fhould live
for fome days, he was fuddenly affected with ftertor, and died.

Appearances on Diffection.

EXTERNALLY. His body had an ugly appearance. It was much emaciated; the thighs were
fcabby; and the abdomen had fallen in.

ABDOMEN. That particular kind of fmell
which generally arifes from inflamed inteftines,
was perceived on opening the belly; and, accordingly, confiderable portions of the fmall inteftines
were in different places found to be of a very red
colour. The edge of the liver, and its hollow furface on the contiguous part, to a fmall extent, was
of a livid colour. The gall bladder had tinged
the pylorus and duodenum very deeply; but the
colour did not penetrate their coats. The flat
                                    furface

furface of the fpleen was very black. The pancreas was a little indurated.

THORAX. The upper part of the right lobe of the lungs was enlarged, and appeared to be much indurated. When cut into, pus, or matter refembling it, together with a frothy fluid, were difcharged; and its fubftance refembled part of the liver. A fmall quantity of greenifh yellow coloured fluid was found in the left fide of the thorax. The left lobe adhered to the pleura towards the diaphragm, and more clofely on the upper fide, through the interpofition of a yellow thin membranous fubftance. The inferior part of the fame lobe was black and indurated, and was of the fame fubftance as that of the right; and like it, too, contained purulent matter of a whitifh colour, which was difcharged when the lobe was drawn out of the thórax. The pleura was quite found. In the pericardium there was a large quantity of greenifh yellow coloured watery fluid. Polypous concretions were feen at each of the orifices of the heart; the fmalleft extended into the left auricle, and the largeft into the aorta. That veffel was much wider than ufual, but had no inequalities. Points of incipient offifications were feen upon its internal furface beyond the valves.

CASE

CASE XLIII. (xxi. 30.)

A MAN, aged fifty-fix years, of a tall ftature, and of a pretty good habit of body, but irregular in his mode of living, having undergone much fatigue in ringing church bells, (the means by which he gained his bread) had complained for fome days of a flight pricking pain within the breaft, at the lower part of the fternum. The pain became much more violent ; and upon the fame day, fever and difficulty of breathing fupervened. He had two or three fpontaneous ftools, confifting of a bilious frothy matter. He was brought into the hofpital of Bologna fo late, that he could not be bled till about the end of the fourth day of the fever. The blood, contained little ferum, and had a whitifh yellow cruft of about two inches in thicknefs. On the fifth day. his pulfe was full and hard : he had fcarcely flept during the night ; and he could not breathe freely. The urine which he paffed during the end of the fifth day was of a deep colour, and not very clear. His pulfe, at the fame time, was frequent, irregular, and not hard. His refpiration was difficult, and attended with moaning. The pain was unabated ; he could lie eafily on either fide ; and he expectorated a thick frothy matter with yellow ftreaks. During the night he was

much

much diftreffed with the pain and cough.   In the
morning the pain was alleviated : but his pulfe
and expectoration were as on the preceding day ;
as was alfo his urine, except that it was paffed in
fmall quantity, and was of a reddifh colour, and
fomewhat turbid.   Towards the end of that day,
which was the fixth of the difeafe, blood was
drawn from his right hand ; but as it was all received into water, it could not be properly examined.   On the beginning of the feventh day he
feemed better ; but towards the end of the fame
day, his pain became aggravated, his pulfe fmaller
and more frequent, his refpiration hurried, and
his tongue parched.   He paffed a reftlefs night.
In the beginning of the fuccceding day he had
profufe fweat, which was encouraged by gentle
means ; but without producing any relief.   Towards the end of that day, his breathing was very
much hurried ; his pulfe fmall, affording little refiftance to the finger when preffed.   He expectorated a yellow matter, which was fluid, and not
frothy ; and his urine continued to have the fame
appearances.   About the beginning of the ninth
day, he could at pleafure put his arms without the
bed-cloaths ; and could alfo fpeak, though with
fome difficulty.   From thefe circumftances, thofe
about him did not imagine that he was dying ;
yet he foon after expired.

VOL. I.   Q q   *Appearances*

Appearances on Diſſection.

ABDOMEN. Externally the belly was livid about the ilia; and under that part the colon, which was diſtended with air, though in every other reſpect ſound, lay. The liver extended ſo much acroſs the body, that it covered the whole upper part of the ſpleen, to which it was cloſe-ly connected. Its edge was for a conſiderable ſpace livid; as was alſo its concave ſurface, to the extent of three fingers breadth : but the lividneſs was quite ſuperficial. It was not harder than uſual, but was a little whiter. The gall-bladder contained very little bile; though the antrum pylori, which lay under it, was tinged with that fluid. It had in it, however, twenty gall-ſtones, of various ſizes, but moſt of them ſmall; except one, which was very large, and which, like the reſt, reſembled charcoal in colour, in roughneſs, and in brittleneſs of ſubſtance. None of theſe gall-ſtones, when expoſed to the fire, emitted flame or ſparks; but they ſometimes crackled a little. The ſpleen was large, even in proportion to the ſtature of the man : it was flabby, and externally of a whitiſh colour. The external ſurface of the ſto-mach, on the whole of the left ſide, was marked with large, and as it were, ramifying ſpots, of a blackiſh livid colour. On the internal ſurface, too, ſimilar ſpots appeared, extending to the œſo-phagus; and about them drops of blood, ſticking

between

between the coats and the ſtomach, were ſeen: ſo that, from every circumſtance, there could be no doubt of the ſtomach having been inflamed.

THORAX. Each ſide of the cheſt contained watery fluid; and that on the right ſide at leaſt was turbid and of a yellow colour. Both lobes of the lungs were at the upper part connected to the pleura; which, there, and in ſome other parts of the right ſide, was conſiderably thickened. On the ſame ſide, pieces apparently of that membrane lay on the ſurface of the lungs, which was in other reſpects ſound. The right lobe of the lungs was very heavy, and its ſubſtance throughout reſembled that of the liver: it was of a whitiſh colour, and indurated, but leſs ſo than uſual under the ſame circumſtances. It appeared ſemiputrid; and more eſpecially, as a whitiſh turbid fluid was in many places diſcharged from the bronchia when cut into. Over that ſurface of the lungs, contiguous to the mediaſtinum, next the pericardium, and over the mediaſtinum itſelf, a thickiſh reticulated ſubſtance, of a yellowiſh colour, and of a beautiful appearance, was extended. It could be eaſily drawn off: this was alſo the caſe with reſpect to another membranous ſubſtance, of a bloody colour, which was alſo ſpread over the mediaſtinum at the ſame part. No ſuch appearance occurred in the left ſide; and on that ſide the lungs were ſound, or nearly ſo. In the pe-

ricardium

ricardium there was a good deal of the fame kind of fluid feen in the right fide of the cheft. The heart was very large, and exceedingly flaccid. It contained two polypous concretions. The one of which beginning in the right auricle, was from thence extended through the adjoining ventricle into the pulmonary artery. The other, which was fomewhat thicker, and much larger, filled almoft the whole of the left ventricle, and from that was continued into the aorta. When it was drawn out of that veffel, a column of ftrongly coagulated blood of the length of a fpan followed. Nor was the blood contained in the veffels above that, lefs coagulated, as appeared when the neck was cut into. The veffels of the larynx and phatynx were turgid. The face was of a livid red colour; and the external ear contained a little half-coagulated blood.

Case XLIV. (xxi. 32.)

A woman, aged forty-five years, affected with an ulcer of long ftanding in one of her legs, with a dry fcabby eruption over her whole body, and with a kind of flow fever, was admitted into the hofpital of incurables at Bologna. Previous to her admiffion, fhe had drank fuch a quantity of wine, that fhe had fcarcely tafted any thing elfe for three days; in confequence of which, fhe was

.very

very hot and reftlefs during the firft night of her refidence in the hofpital. Next morning fhe was better; but her pulfe was frequent, quick, hard, and cord-like, though it did not refift much when preffed by the fingers, nor yet was it full. In the evening, fhe again became hot, and had felt an acute pricking pain at the lowermoft true rib on the left fide. The pain was aggravated neither by external preffure, nor by lying upon that fide. She lay, however, more eafily on the right fide; for, when on the left, the cough which in this difeafe always attends, was excited. The pain afterwards ceafed, and did not again recur; fo that fhe feemed to be fomewhat better, more e-fpecially as the pulfe, though in other refpects the fame, had become lefs frequent. But on the fourth day, after a fhivering fit, fucceeded by a hot one, fhe became worfe; and along with her cough, which had hitherto been dry, fhe expec-torated a bloody matter, of a cineritious livid co-lour, and of a fetid fmell. She could not breathe unlefs her neck were erect. Her pulfe became fmaller and weaker, and fhe expectorated black coloured purulent matter. A flight delirium fu-pervened. The exacerbations of the fever occur-red at a later hour on thefe latter days. In the middle of the fixth day, fhe died as if fhe had been fuddenly fuffocated.

Appearances

Appearances on Diffection.

ABDOMEN.   All the vifcera were found.

THORAX.   Each fide of the cheft contained a fmall quantity of fetid ferous fluid, of a cineritious colour.   The lungs adhered very flightly to the pleura.   That membrane, on the left fide, where the lungs had adhered, was in different places rough and unequal, in confequence of fmall red coloured tubercles.   Adjoining to the left lobe of the lungs, at that part where it is contiguous to the diaphragm, there were two falfe membranes; one of which adhered to the lobe itfelf, and the other to the diaphragm.   That lobe was not red, but of a livid colour, and was of a harder confiftence than the liver.   When cut into, matter like what had been expectorated was found in its internal fubftance, and efpecially in certain finufes, as it were, which contained yellow pus.   No hard nor particular coat invefted thofe finufes.   There was nothing within the afpera arteria.   The other lobe was quite found.   Small polypous concretions were obferved in the vena cava and in the pulmonary artery; and a pretty large one in the left auricle.

CASE XLV.   (XXI. 3.)

A ROBUST young man, about eighteen years of
age,

age, affected with pneumonic inflammation, died within eight days.

Appearances on Diffection.

ABDOMEN. Some effufed ferous fluid was feen in the abdominal cavity. The edge of the liver was livid. The inteftines were in fome places of a reddifh colour, and had a ftrong fmell.

THORAX. The right lobe of the lungs adhered ftrongly everywhere to the neighbouring parts, by means of a thin membranous fubftance. The upper part of the fame lobe was much indurated, and very heavy; its fubftance refembling that of the liver: an appearance which took place in the remaining part of the fame lobe, and throughout the greateft part of the left. The left fide of the thorax contained a large quantity of bloody, black-coloured, watery fluid; and a good deal of fimilar fluid was found in the pericardium. The right auricle of the heart was very much dilated, probably in confequence of its having eafily yielded to a quantity of blood thrown into it during the laft moments of life. The blood was black and grumous, and furrounded a large firm polypous concretion, part of which belonged to the adjoining ventricle. Another fubftance, of the fame kind, but of a round form, appeared in the pulmonary artery. The left ventricle and auricle contained no fuch fubftance; nor had there ever been any

blood

blood in them, except it had flowed out during the diffection.

CASE XLVI. (xx. 17.)

A YOUNG man, almoft thirty years of age, after much working, became affected with a flight dull pain in the left fide of the thorax. At the fame time he was feverifh, he breathed with difficulty, and had no expectoration. Thefe fymptoms continued for fourteen days, about the end of which time they feemed fomewhat alleviated. But, on a fudden, great difficulty of breathing, attended with ftertor, and the expectoration of a rofy coloured matter, which he fpit up in great quantity and with no difficulty, fupervened. He had lain during the whole courfe of the difeafe upon his right fide, and fometimes upon his back. On the feventeenth day, while turning upon the right fide, he expired.

Appearances on Diffection.

THORAX. The pleura appeared found, and totally unconnected with the lungs. The left lobe of the lungs was univerfally inflamed, except at the fuperior part. From that part, although found, blood flowed in drops at two places; fo that more than four pounds of that fluid were found ftagnating in the left fide of the cheft. The heart contained two polypous concretions.

3

CASE

CASE XLVII. (LVIII. 13.)

A BUTCHER, aged about forty years, affected with lues venerea, often intoxicated, and fo fub-ject to difeafes of the thorax, that he had been re-peatedly in the hofpital of Padua, on account of fuch complaints, was again admitted into that hof-pital. He laboured under acute fever; and was troubled with conftant cough, which was fo much aggravated three or four times every hour, that he became livid from the exertion. He had alfo pu-rulent expectoration, and a hard chord-like pulfe. Venefection was twice performed, and the blood at both times had a buffy coat. Having been thus affected for fifteen days, at laft, within the fpace of a fingle day, his ftrength failed more and more, and he died.

Appearances on Diffection.

THORAX. The lungs were quite putrid, and had a very offenfive fmell. The heart was flabby. In one of the valves of the aorta, the Papilla A-rantii was much larger than natural. The mem-branous layers of which the valve is compofed, at the furface, under the above-mentioned papilla, that is, oppofite to the other valves, were feparated from each other to a confiderable extent, fo that a probe could be introduced into the opening. The contiguous portion of the trunk of the aorta was

VOL. I.   R r   marked

marked internally with white fpots, and was fomewhat unequal on its internal furface. The aorta,
at its arch, was diftended into an aneurifm.

C A S E XLIII. (xx. 43.)

An unmarried woman, of twenty-two years of
age, was feized with a fixed acute pricking pain in the
right fide, attended with cough, expectoration, difficulty of breathing, and fever. She could lie in no
fituation, but upon the pained fide. In the progrefs
of the difeafe, her expectoration became tinged
with blood, and one day a large quantity of that
fluid was fpit up. For fome days before death
fhe was affected with pains about the ilia. On
the ninth day, convulfive motions having fupervened, fhe died while lying on her left fide.

Appearances on Diffection.

Abdomen. Some ferous fluid was feen in the
belly. The fpleen was very large. The greater
part of the fmall inteftines were inflamed.

Thorax. The right lobe of the lungs was
ftrongly connected to the pleura. Its whole fubftance was greatly inflamed, and in one part next
the pleura an ulcer was obferved; between which
and the pleura fome ferous fluid appeared. The
left lobe approached towards a black colour;
and was alfo marked here and there with black
fpots. The left fide of the thorax contained
fome

fome ounces of ferous fluid. Two polypous concretions were feen in the heart; the fmaller lay in the
left ventricle; and the larger filled the whole of
the right auricle, which was very much dilated.

Case XLIX. (xx. 45.)

A man, aged fifty years, was feized with an
acute fever, laborious refpiration, pain, extending
from the fternum almoft to the abdominal mufcles, and moft diftreffing cough, which occafioned great pain about the left breaft and falfe ribs.
He could lie on neither fide. He had no expectoration. On the fifth day of the difeafe he died.
Appearances on Diffection.

Thorax. When the fternum was raifed, a fmall
quantity of dark-coloured ferous fluid flowed out
from the left fide of the cheft. The left lobe of
the lungs was much indurated, approached towards a greenifh colour, and was marked with
black fpots. When cut into, a watery and putrid colluvies, with blood intermixed, was difcharged. Polypous concretions were found on the
ventricles of the heart.

Case L. (xx. 41.)

A woman, of fixty years of age, was feized
with an acute pricking pain in the right fide of

the

the thorax, fo that she could not lie on that side. She was troubled with cough, attended with little expectoration.   She breathed uneasily, but not with exceffive difficulty; and her ftrength was entirely exhaufted.   The pain afterwards became fo much alleviated, that fhe could lie eafily on the affected fide.   Her pulfe however having every day grown weaker, fhe at length died.

Appearances on Diffection.

ABDOMEN. Within the membrane of the fpleen, an offeous body, of a fpherical form, was found.

THORAX. Both lobes of the lungs adhered to the pleura.   The right lobe, which adhered more clofely than the left, was at the fuperior part much indurated.   In the fame part an abfcefs was obferved : this, when cut into, difcharged a large quantity of dark-coloured ferous fluid.   The contiguous fubftance of the lungs was tinged with the fame colour, and feemed in fome meafure affected with gangrene.   The left lobe, which did not like the right confift of a fingle portion, but was compofed of feveral lobules, was flightly inflamed at the pofterior part, and throughout the whole external furface was here and there marked with black fpots.   Much watery fluid was found in the pericardium.   Both ventricles of the heart contained a large polypous concretion; that in the right was the largeft.   The blood in this fubject was almoft completely coagulated.

CAUSES

CAUSES of PNEUMONIC INFLAMMATION.

PREDISPONENT CAUSE. There is a natural and
acquired predifpofition to pneumonic inflammation. Under the former head, may be claffed the
general difpofition to inflammatory complaints and
ftraitnefs of the cheft. Under the latter, certain
occupations in life, as the dreffing of flax, fifting
of wheat, chiffeling of ftones, &c. and former inflammatory affections of the lungs.

EXCITING CAUSES. Every circumftance which
is an exciting caufe of general inflammation, may,
in a perfon predifpofed to it, induce pneumonia;
fuch as, expofure to cold, &c. Violent or long
continued exertion of the lungs, too, whether in
fpeaking, finging, or blowing on mufical inftru-
ments, is found to be productive of the fame effect where there is a general inflammatory predifpofition; even although no caufe, directing the inflammation immediately to the lungs, fhall concur. Cutaneous eruptions alfo, particularly meafles, often prove exciting caufes of this difeafe.

Befides the exciting caufes now enumerated,
MORGAGNI has mentioned fome circumftances,
(Epiftle xxi. N° 43. & 44.) which feem to fhow,
that pneumonia may arife from the prefence of

worms

worms in the inteftinal canal. He ftates particularly, that pneumonic inflammation, attended
with worms, was very prevalent at Farnefe in the
winter of 1705 *. But as the difeafe was not
cured

---

* Dr. Pedratti, who then practifed at Farnefe, gave MORGAG-
NI the following acconnt of the fymptoms which characterized
that epidemic. The firft fymptom was pain in the fide, rather
obtufe than acute; at the beginning tolerable, and throughout its
courfe fometimes intermitting. In the mean time, figns of worms
harbouring in the alimentary canal appeared; and befides, fome of
thofe animals were thrown off by vomiting, and fome of them
were feen in the fæces. At the fame time there was violent cough,
attended with the expectoration of a white crude matter tinged
with ftreaks of blood. Fever, ufhered in by fhivering, continuing
uniformly without either remiffion or exacerbation, accompanied
thefe fymptoms. The pulfe was low, fmall, and unequal. About
the fifth, or at the utmoft the feventh day, the difeafe, inftead of
being aggravated, feemed rather to decreafe in violence: fo that
the patient appeared in a way of recovery, the pain and cough
having ceafed, and the fever even being much moderated. Soon
after this, however, all thefe fymptoms became more violent than
before; with the addition of fo great a difficulty of breathing, and
fuch proftration of ftrength, that death, preceded by burning heat
in the internal parts, while the external furface of the body was
cold, and, what was the moft certain fign of approaching diffolution, univerfally livid, like that of a dead body, took place within forty hours. Vomiting, at the beginning of the difeafe, efpecially when excited by the preparation called Ruland's Water,
was found ufeful in robuft patients. But purging, even by means
of frefh drawn almond oil, or by calomel combined with a little
myrrh and coralline, did not produce fuch good effects. For although, by the latter medicine, the worms, which were of the
lumbricus fpecies, were in a wonderful manner expelled; yet immediately after, the ftools, the pain was aggravated, and the inflammation

cured by the expulfion of the worms, and as the appearances on diffection in thofe who died of that diforder were pretty nearly the fame as in other cafes; it is more probable that the prefence of the worms was an accidental circumftance only, than that it could be an exciting caufe of the difeafe.

—————

REMARKS on the Cases of Pneumonic Inflammation.

In thefe cafes, the various degrees of the feveral fatal terminations of pneumonic inflammation are very accurately pointed out. Thus, in the firft ten cafes, inflammation and induration of the fub-ftance of the furface of the lungs are exhibited. In the two following ones, the pericardium and pleura invefting the ribs had alfo been inflamed, and the

flammation feemed increafed. On the other hand, by the ufe of the oil, the expectoration appeared at firft to be rendered more eafy; but foon afterwards, as if the lungs had been thereby re-laxed, the difficulty of breathing became greater, and the death of the patient was accelerated. For thefe reafons, it was neceffa-ry to lay afide the employment of thofe medicines. Befides, as blood-letting was found to deprefs the ftrength, it could not be had recourfe to, except in cafes where the i flammatory fymp-toms were exceedingly urgent, and then a fmall quantity only was taken away at a time.

the inflammation had there terminated in effufion. From the fourteenth to the twenty-fourth,
both inclufive, the various gradations in effufions
within the fubftance of the lungs, from bloody ferum to perfect pus, are well marked. The fucceeding cafes as far as the thirty-eighth, are inftances of empyema exifting, either fimply or combined, with exudation from the furface of the pericardium. Empyema and vomica were joined
in the following eight cafes. The forty-fifth and
forty-fixth, are examples of bloody extravafations
within the thorax; and the four laft cafes feem to
fhew the progreffive ftages of gangrene.

The obfervations already offered refpecting the
phenomena of inflammation, render it unneceffary
to introduce any extended remarks in this place.

The hiftories of the firft ten cafes prove, that
a degree of inflammation in the lungs, which
in many other parts could not be productive of
much injury to the fyftem, occafions fatal event.

The induration of the fubftance of thofe organs, it has been already alleged, is owing to many of the veffels diftributed over the air cells being diftended with blood: but it is probable that
fome other circumftance concurs; for in that ftate
their fubftance is as compact and heavy as that
of the liver, and, like it too, exhibits a fmooth
fhining furface when cut into.

In the firft cafe, although delirium had taken

place,

place, there was no inflammation within the cranium.

The feventeenth cafe affords an inftance of the general fatality of acute difeafes during pregnancy. Under fuch circumftances it commonly happens that abortion precedes the fatal event : but this furnifhes an exception to the general rule.

Cafe eighteenth contains the hiftory of a patient who died from pneumonia, which was in the winter of 1738 epidemic at Padua, efpecially in fome convents of nuns. MORGAGNI feems to hint that the difeafe was deemed contagious : but he afferts, as a proof of the contrary, not only that it was not communicated to the attendants of the fick, but alfo that all who were affected had a previous difpofition to pneumonic inflammation.

The twenty-fifth cafe is an example of cynanche terminating in pneumonia.

The fymptoms of the thirty-firft cafe refembled fo much thofe of inflammation of the liver, that Valfalva, after confiderable hefitation, concluded it to be that difeafe. He was led to form this conclufion from there being no fymptoms diftinctly characterifing pneumonia, and from the patient conftantly pointing with his own hand to the region of the liver as the feat of the pain. It muft be allowed that every fymptom of hepatitis was prefent, except the pain in the clavicle or fhoulder, which alone in this inftance might have fhewn that there

VOL. I.   S s   was

was no inflammation of the liver. At the fame time cafes of pneumonia may occur, attended even by this fymptom. Fortunately the practice in both cafes, being nearly the fame, is not very materially influenced by the diftinction of the difeafes.

In feven cafes, viz. the eighth, nineteenth, twenty-fourth, forty-fecond, forty-third, forty-fifth, and forty-eighth, fome of the contents of the abdomen were inflamed. Whether this is to be regarded as an accidental circumftance, or as the confequence of fympathy, remains to be determined.

SECT. IV. *HEPATITIS;* or, Inflammation of the Liver.

HEPATITIS is faid to be characterifed by pain and tenfion in the region of the liver, the pain extending to the right clavicle and top of the fhoulder; difficulty of lying on the left fide, breathlefsnefs, dry cough, vomiting, hiccup, and fymptoms of inflammatory fever *.

In

---

* Vid. Cullen Nofolog. Method. edition quart. vol. ii. p. 113.

In temperate climates this difeafe occurs rarely; but the confequences of inflammation, as fcirrhous hardnefs, fuppurations, &c. are often obferved in the liver after death, when no fymptom indicating inflammation of that organ had preceded. Such cafes have been termed Chronic Hepatitis : a title which appears exceedingly improper.

The fymptoms of true hepatitis are different in different cafes.

Sometimes the pain refembles, both in the fenfation it communicates, and in its feat, that of pneumonia, and, like it too, is increafed by refpiration; but it is different in one refpect, viz. in being aggravated by the touch. In fuch cafes, the feat of the inflammation is found to be on the convex furface. In other cafes, the pain is attended with vomiting, great anxiety, and jaundice. Under thefe circumftances, the concave furface of the liver is chiefly affected.

Obftinate conftipation is faid to attend at the beginning, in many cafes. In others, the fæces paffed have a white appearance.

This difeafe is by no means fo dangerous as the inflammation of other abdominal vifcera : for it often ceafes fpontaneoufly, terminating by a critical difcharge of urine, fweat, bilious ftools, or hæmorrhagy from the nofe or feat of the piles.

Where it ends fatally, it feldom proves rapidly or fuddenly fatal; probably never, unlefs where

the

the inflammation is communicated to the ſtomach
or bowels.   It moſt generally terminates in ſuppuration and abſceſs, from which the matter is forced into the abdomen or thorax ; and hence the
patient either becomes hectic, or is ſuddenly ſuffocated.

But even when ſuppuration enſues death is not
always the conſequence : for ſometimes the matter is by the biliary ducts conveyed into the inteſtinal canal, and ſo thrown off ; ſometimes it is diſcharged by an external opening through the parietes of the abdomen ; and ſometimes too it is
coughed up from the lungs.   Gangrene ſeldom
takes place.

The ſeat of the diſeaſe is probably firſt in the
external membrane of the liver, from which it
is communicated 'to the parenchymatous ſubſtance.   Some authors have ſuppoſed that, in true
hepatitis, the inflammation is confined to the former of theſe parts ; and in what they have ſtyled
the chronic hepatitis, it is excluſively ſeated in the
latter *.   But this opinion is not confirmed by
the appearances on diſſection.

CASE

---

* See Dr. Cullen's Firſt Lines, par. 415.

CASE of HEPATITIS.

(XXXVI. 4.)

An old woman, more than fixty years of age, had long complained of a pain above the umbilical region, attended with thirft, cough, and the expectoration of a catarrhous matter. Towards the end of her life her refpiration was difficult ; and a few days before death, her belly became fuddenly much fwelled, and her feet œdematous. At length, the pain above the umbilicus having gradually abated, fhe died.

Appearances on Diffection.

ABDOMEN. The belly contained a great quantity of limpid watery fluid. No veftiges of lymphatics were perceived. The fpleen was twice the ordinary fize. The liver was indurated. A congeries of veficles, from which ferum was difcharged when they were ruptured, appeared adhering to one part of the liver. Within its fubftance, next the diaphragm, the cavity of an abfcefs, occupying more than one third of the organ, was found. From this the matter had burft through the diaphragm into the right cavity of the cheft, which was filled with fanious matter. The included lobe of the lungs, however, was

3   .   found.

found.   The gall-bladder was full of fmooth calculi.

CAUSES of Hepatitis.

Predisponent Cause.   The moft probable
conjecture on this fubject, is, that the ufe of particular kinds of food predifpofes to this difeafe.

Exciting Causes.   All the general exciting
caufes of inflammation may induce hepatitis.   Of
this the moft frequent is external violence, and
more efpecially, it has been obferved, if that accident have occafioned fracture of the fkull.   It
has been alfo alleged, that certain paffions of the
mind, violent fummer heats, violent exercife, intermittent and remittent fevers, folid concretions
or collections of fluids in the liver produced by unknown caufes, and chronic inflammation of that
organ, prove exciting caufes of the difeafe *.

REMARKS on the Case of Hepatitis.

Morgagni regrets that the fymptoms of the
cafe were not more accurately marked by Valfalva.

* Vide Dr. Cullen's Firft Lines, par. 416.

va. The author of thefe remarks has ranked it as an example of hepatitis, from confidering not only the appearances on diffection, but alfo the expreffion in the hiftory of the cafe, that " the pain gradually abated." This, in his opinion renders it probable that the pain had been a principal fymptom.

Although the morbid appearances be not detailed fo accurately as might have been wifhed, there is reafon to conclude that both the external membrane of the liver, and the parenchymatous fubftance had been inflamed. It cannot however be determined, whether the inflammation had originally begun in the former or in the latter part.

The exciting caufe was probably the congeries of veficles attached to the external furface of the liver.

SECT. V. *GASTRITIS;* or, INFLAMMATION OF THE STOMACH.

THE fymptoms which characterife this difeafe are by no means accurately pointed out by authors; a circumftance that may be attributed to two caufes, The infrequency of the cafe, and the

extenfive

extensive nervous influence of the affected organ. While the former of these causes prevents any individual practitioner from collecting such a number of observations as entitles him to draw general conclusions; the latter must vary the type of the disease, in different cases, according to the susceptibility of impression of the nervous system in the person affected.

The symptoms commonly enumerated * are, fever, with small, quick, hard pulse; violent burning pain in the region of the stomach; painful vomiting, more especially after any thing is swallowed, whether it be liquid or solid; most distressing thirst; constant watching; great restlessness, and continual tossing of the body; hiccup; difficult respiration; coldness of the extremities; fainting; intermitting pulse; and convulsions.

In addition to these symptoms, Dr. Cullen has remarked that gastritis is attended with greater loss of strength than any other inflammatory disease; and Meza has mentioned, that suppression of urine, without any distension of the bladder, or, in other words, the suspension of that secretion,

takes

---

* Vide Translation of Hoffman, revised by Dr Duncan, vol. i. p. 229. Macbride's Works, 4to edit. p. 449. Meza, loco citato, p. 27. Joseph Quarin, loco citato, p. 351. Dr Cullen's First Lines, vol. i. p. 412.

takes place. In the following cafes, purging as well as vomiting occurred.

The pathognomonic fymptoms are faid to be, violent burning pain, with fwelling about the fto-mach, painful vomiting and eructation, and hic-cough. But a cafe is noticed by De Haen [*], where not only was the vomiting abfent after the firft attack, but alfo did the appetite for food con-tinue unimpaired. On the other hand, Sauvages obferves [†], that a difeafe prevailed at Montpelier, in the fummer of 1760, which refembled gaftritis in every fymptom, but that of vomiting. He imagined that the fterno-coftalis mufcle had in thefe cafes been inflamed.

The event of this difeafe is feldom favourable, as it moft generally terminates in fuppuration and ulceration, or in gangrene. Death, too, fometimes happens merely from the violence of the inflam-mation, before either fuppuration or gangrene can take place. This can only be accounted for from the wonderful connection between the vafcular and nervous fyftem, and the great nervous influ-ence which the ftomach poffeffes.

The feat of the difeafe has been faid to be ei-ther in the villous coat of the ftomach, or in its nervous coat, as it is termed, and the peritoneum

Vol. I. T t inveiting

* Ratio Medendi, vol. iii. p. 30.
† Nofolog. Method, tom. i. p. 478.

invefting it. In the former cafe, the inflammation is of the eryfipelatous kind; in the latter, of the phlegmonic. The latter alone is attended with the fymptoms conftituting gaftritis. The former is commonly a fymptomatic difeafe; and is not accompanied by marks of general inflammatory affection, nor by acute burning pain in the ftomach.

CASES OF GASTRITIS.

CASE I. (XXX. 4.)

A NOBLEMAN, aged forty-two years, became affected with a double tertian fever, which, during the firft paroxyfms, was accompanied with mild fymptoms. On the fourth acceffion, however, the fymptoms were exceedingly violent, for the cold fit continued feven hours. He had very troublefome thirft; his tongue was rough; his breathing difficult. He was languid; his pulfe was fmall and weak; and he complained of pain, and fenfe of fulnefs in the ftomach. He was at the fame time fo uneafy and reftlefs, that he could not continue even for a little time in the fame part of the bed. All thefe fymptoms became milder when the hot fit took place, and when he was allowed to drink diftilled water. But this relief lafted on-

ly

ly a fhort time; for an aggravation of his complaints foon returned, and he continued much diftreffed during the whole of that night. Early in
the morning he felt an inclination to vomit,
which at firft could not be excited even by the
irritation of the fauces by his fingers, which he
did feveral times. Soon after, however, he vomited about four pounds of fluid, refembling in
colour water in which chocolate had been diffolved. In this fluid, membranous-like fubftances,
of the fame colour, floated; and from it the fame
kind of fmell ufually emitted from the bodies of
thofe affected with fever proceeded. Although
the affection of the ftomach abated fomewhat after the vomiting; the other fymptoms not only
continued, but were even aggravated. In the
morning venefection was had recourfe to. The
blood drawn in the firft cup had its craffamentum
fofter than ufual; had a thin cruft on its furface,
and exhibited a milky-like ferum: but in the fecond cup, the blood was much lefs altered. Notwithftanding the employment of thefe and other
means, almoft the fame quantity of the fame kind
of fluid as that formerly mentioned was vomited
within a few hours; and this a little after happened again and again: fo that the whole quantity
vomited during that day was equal to fixteen
pounds. On the following night all the fymptoms
were violent; and befides, a tremor of the left

T t 2   arm

arm fupervened. This recurred frequently, efpecially when the arm was expofed to the air; and
was always preceded by delirium. In the morning this tremor was converted into a kind of epileptic paroxyfm; by which not only his arm, but
alfo his mouth, eyes, and left thigh, were violently convulfed. Thefe fymptoms continued for
many hours; and at length the affected arm became paralytic. The epileptic fits recurred fo
frequently, that more than twenty took place
within an hour. At the fame time, too, the vomiting became more frequent; and the matter
thrown up was of a porraceous colour, and had
fmall membranous portions floating in it. Sin-
gultus, which had begun about mid-day, (after the
paralytic attack) now became more violent. Although all thefe fymptoms feemed fomewhat a-
bated in the afternoon, on the approach of evening they were aggravated; and the pulfe and
ftrength having failed more and more during the
whole of the night, while he continued diftreffed
with vomiting, delirium, fingultus, and frequent
violent, though fhort, fpafmodic attacks, he died in the morning.

Appearances on Diffection.

ABDOMEN. The belly and the inteftines were
fwelled. The anterior furface of the ftomach and
inteftines were tinged with the fame colour as the
matter which had been vomited. Internally the
ftomach

ftomach was inflamed, and all its moft minute veffels were very much diftended with blood. The
gall-bladder, although it contained no bile, was
turgid from air.

THORAX. The right lobe of the lungs adhered
ftrongly to the pleura: both it and the left were
tinged with a black colour, and were full of ichorous matter. In the right ventricle of the heart
there was a thin polypous concretion.

C A S E II. (XXIX. 18.)

A MAN, aged forty years, of a brawny make,
and accuftomed to intenfe thought, having been
affected for fome days with pain in the head, and
a fenfe of heat in the parts on making water, was
feized, after fupper, at which he neither eat too
much, nor any thing unwholefome, with excruciating pains in the region of the ftomach. The
pain of the head continued, and the pains in the
ftomach became aggravated. He difcharged, by
the mouth and by ftool, a large quantity of green-
coloured mattter. Under thefe circumftances, he
died at the beginning of the third day.

Appearances on Diffection.

ABDOMEN. The right fide of the ftomach was
found; and on its internal furface, about the
antrum pylori, to the extent of feveral fingers
breadth, numerous lenticular glands, much enlarged,

larged, appeared. The fundus, on the left fide, was maiked with bright-coloured bloody fpots. Some of thefe fpots having begun to be covered with an ugly ferruginous incruftation, fhowed that the inflammation had approached to the ftate of gangrene. On the fame fide, at the part on which there were no fpots, the internal coats feemed found ; but difcharged blood very readily when preffed. The duodenum, and reft of the inteftinal canal, exhibited no morbid appearance. The gall-bladder, at the diftance of two or three inches from the lower part of its fundus, was contracted, and became again dilated before it terminated in the cyftic duct : fo that it appeared to form a double gall-bladder.

THORAX. The lobes adhered, by means of their own membrane, to the parietes of the cheft, and alfo to the mediaftinum : they were however found, although their pofterior part was red. The back and pofterior parts of the arms were alfo of a red colour. The heart contained no blood, either within the ventricles or auricles.

CASE III. (XXIX. 20.)

A POOR country woman, apparently about fifty years of age, was fubject at times to difficulty of breathing, together with a fenfe of ftraitnefs in the cheft ; accompanied with hard pulfe, and with

fo

fo great pulſation in all the arteries, that their alternate action could be diſtinctly ſeen both in the
hands, in the neck, and in the temples. When
diſtreſſed with the difficulty of reſpiration, ſhe was
accuſtomed to come to the hoſpital of Bologna,
where ſhe was relieved by being bled freely. The
blood drawn at theſe times was ſomewhat hard.
In this manner ſhe lived for four years; when at
laſt having at her own home became affected with
pains in the ſtomach, ſhe died within twenty-four
hours.

Appearances on Diſſection.

ABDOMEN. The ſtomach was large, and half
full; but when opened, it ſeemed aſtoniſhing that
all its contents had not been evacuated by vomiting. For internally it appeared eroded in ſeveral
different places. The eroſions, although evidently recent, were already affected with a gangrenous colour. They were exceedingly numerous,
and very minute about the pylorus; from which
they alſo extended to the contiguous part of the
duodenum. Eroſions, of a larger ſize, were ſeen
here and there on the fundus, about the cardia,
and even in the œſophagus. Theſe eroſions, therefore, ſeemed to have been produced by what had
been ſwallowed: but what that was, could not be
learned, neither from the previous hiſtory of the
caſe, nor from an examination of the matter that
remained in the ſtomach. The ſpleen was ſomewhat

what larger and more flabby than ufual; and was glued to the diaphragm for a confiderable extent, and alfo to the ftomach in a fmall degree: this might perhaps arife from its large fize. The uterus lay very much to the left fide; and on that fide the round ligament was fhorter than on the right. Within the cervix uteri, at one fide, a fmall thick white membrane, of a pyramidal form, adhered, by a flattened head, which feemed to be the remains of a hydatid formerly diftended with fluid. The fanguiferous veffels of the urinary bladder, even from the openings of the ureters, were fo exccedingly red, that, although they were very fmall, they could not have been better feen, nor could their anaftomofes have been more diftinctly marked, had they been filled with wax. In this manner they were continued from both fides into the urethra; along the internal furface of which they were obferved in great numbers, and much diftended; but, on that account, they did not form fo beautiful an appearance as in the bladder. On cutting into the mefentery, which was well fupplied with fat of a good colour, (as the other parts of the body were, and in a greater degree than could have been fuppofed from the appearance of the fubject) feveral large glands were obferved. Thefe were found; but fome of them equalled in fize the largeft bean. The beginning of the fuperior mefenteric artery was com-

mon

/

ſion alſo to the celiac. The coronary artery of
the ſtomach was much larger than natural. The
vena cava, though cut through both above and
below the liver, did not diſcharge a ſingle drop of
blood.

THORAX. Both ſides of the cheſt contained a
ſmall quantity of colourleſs watery fluid. The
lungs were turgid with air, and adhered to the
pleura at the back and at the ſides. Some polypous concretions appeared in the heart; and alſo
in the jugular veins, which contained more blood
than the veins below them did. The parietes of
the heart were obviouſly thicker than natural on
the left ſide, and thinner than uſual on the right.
There was not, however, any dilatation of the
ventricles, nor of the pulmonary artery and veins,
nor of the aorta. The valves at the mouth of the
aorta were ſomewhat hard. Within the trunk of
that veſſel, both near the heart and at other places, here and there, ſpots appeared, marked only
by a yellow colour, which were probably the beginnings of future oſſifications; for a little above the
diaphragm, where they were of a larger ſize, and
more protuberant, they were already indurated.
The diameter, however, of that veſſel, was in no
part enlarged. This was not the caſe with all its
branches; for, beſides the coronary artery already
mentioned, the left carotid was larger than the
right. That veſſel was divided into two branches

VOL. I.   U u   within

within an inch and an half of its origin, which is very unufual; and, at its divifion, was more dilated than arteries generally are, where they fend off branches. The fame circumftance appeared in the firft divifion of the fubclavians.

HEAD. Although the brain was not examined till the twenty-eighth day after death, it not only appeared found, but alfo had no bad fmell.

CAUSES OF GASTRITIS.

PREDISPONENT CAUSE. It cannot be doubted that fome particular ftate of the ftomach predifpofes to gaftritis; but its nature is fo obfcure, that any attempt to inveftigate it would be exceedingly unfatisfactory.

EXCITING CAUSES. Befides the ordinary general exciting caufes of inflammation, every acrid fubftance applied to the ftomach, whether in the form of aliment, medicine, or vitiated bile, or mucus, has been known to produce gaftritis. Thus, it has been excited by external injuries, expofure to cold, or cold and wet; the receffion of cutaneous eruptions, or the fuppreffion of habitual evacuations; by the ufe of fpirituous liquors, exceffive indulgence in eating, and draughts of cold water when the body is heated; by fwallowing acids

acids and alkalis, or poifonous fubftances, fuch as arfenic, &c. and thofe various purgative medicines which produce their effect principally by ftimulating the mufcular fibres of the inteftines; and by an over-proportion of mucus or bile, or a vitiated ftate of thefe fecretions. The inflammation is fometimes too communicated from neighbouring parts.

Some of thefe circumftances occafion gaftritis in every different ftate of the fyftem: fuch, for example, are certain external injuries, fwallowing concentrated acids, or alkalis, or arfenic, &c. But others, as all the general exciting caufes of inflammation, the ordinary laxative medicines, over-proportion of mucus, bile, &c. are productive of no fuch effects, unlefs fome change from the healthy ftate, either in the ftomach or fyftem, or both, have previoufly taken place. The nature, however, of that change, as has been already mentioned, is ftill involved in obfcurity.

REMARKS ON THE CASES OF GASTRITIS.

THE exciting caufe in all thefe cafes is obfcure. In the firft cafe, indeed, it probably was an over-proportion of bile in the ftomach; which is a frequent confequence of intermittent fever. In the fecond, the patient had been habituated to in-

tenfe

tenfe thought, which is well known to interrupt
the functions of the ftomach : hence an accumulation of mucus, or fome morbid alteration in the
food, might have taken place. In both inftances
fome particular circumftance muft have predifpofed to the difeafe ; for it is not an ordinary effect,
either of intermittents or of intenfe thinking.

The third cafe is ftill more obfcure than the
former. The erofion in the œfophagus, as well
as in the ftomach, induced MORGAGNI to think
that the difeafe originated from fomething fwallowed.

SECT. VI. *ENTERITIS;* OR, INFLAMMATION
OF THE INTESTINES.

INFLAMMATION of the inteftines is marked
by fixed pain, with tenfion in fome part of the
belly, efpecially about the umbilicus ; naufea, vomiting ; obftinate conftipation, attended with low,
quick, hard pulfe ; great thirft, dry tongue, burning heat, and red-coloured urine.

The pain fometimes invades fuddenly, and fometimes is uſhered in by ſhivering, or diarrhœa. Although fixed, it is occafionally much aggravated,
but

but is never entirely alleviated. It is increafed upon touching the affected part; and is commonly
accompanied by hardnefs and tenfion of the abdomen. The naufea and vomiting are more fevere in many cafes than in others; which has been
thought to depend upon the vicinity of the inflamed part to the ftomach. Apparent diarrhœa
fometimes attends; hiccup often. Sometimes this
difeafe imitates, to a fuperficial obferver, hepatitis and gaftritis : but from the former it is diftinguifhed by the abfence of the pain on the top of
the fhoulder, and from the latter by the painful
fenfation about the umbilicus. In fome cafes,
however, all thefe difeafes are combined. On
fome occafions, too, the fymptoms of inflammation of the inteftines are fo obfcurely marked, that
they may be entirely overlooked. MORGAGNI
has mentioned, on the authority of Valfalva, that
where there is hardnefs and tenfion, with a flight
degree of pain in the abdomen, attended with low,
weak, unequal pulfe, and an unufual appearance
in the face, from wildnefs in the eyes, or livid-
nefs round the mouth, there is always reafon to
fufpect inflammation of the inteftines.

This difeafe is a very dangerous one, as it fometimes terminates in gangrene within a few hours
from its commencement, and often within a day
or two. Sometimes, too, it proves fatal during
the inflammatory ftage. It terminates alfo by
exudation

exudation into the cavity of the abdomen or of the inteſtines; by ſuppuration, and ulceration, and ſometimes by reſolution.

The ſymptoms which indicate gangrene having taken place, are, ſudden ceſſation of pain, while the pulſe continues quick, ſmall, and unequal, and at the ſame time the extremities are cold, and the countenance remarkably dejeċted; the vomiting of fæces, or of the glyſters that have been exhibited; and the diſcharge of black fetid ſtools.

Fatal event during the inflammatory ſtage may be dreaded, if, along with ſuppreſſion of urine, violent hiccup, delirium, and ſubſultus tendinum, or convulſions, ſupervene.

Exudation into the abdomen is marked by ſudden ſwelling of the belly, and evident fluctuation in that cavity. When the exudation is into the inteſtines, it is known by the appearance of the ſtools.

Suppuration and ulceration are announced by conſtant vomiting, burning pain, and great ſwelling of the belly.

If the pains be gradually alleviated; if natural ſtools be paſſed; if univerſal ſweat, attended with firm equal pulſe, or if a copious diſcharge of loaded urine, with the ſame ſtate of the pulſe, take place, reſolution may be expeċted.

The ſeat of the diſeaſe is ſometimes in the large inteſtines; more often in the ſmall. Both coats
of

of the inteſtines are probably affected in real enteritis; for there is reaſon to believe, that when the
internal coat alone is inflamed, the ſymptoms e-
numerated do not occur. In ſome caſes, however,
the inflammation is communicated from the internal to the external coat, and *vice verſa.*

CASES of ENTERITIS.

C A S E I. (xxxv. 2.)

A STUDENT of medicine, fond of ſolitude, and
naturally eaſily iraſcible, when, as uſual, in perfect health, became, without any apparent exciting cauſe, except that he knew his father was then
at the point of death, and expected every hour to
receive intelligence of that event, ſuddenly affected, about the fourth or fifth hour of the night, with
a violent pain in the umbilical region, more exceſſive ſometimes in one part, and ſometimes in
another, but never leaving a certain ſpace of that
region. A phyſician having been called, preſcri-
bed a doſe of the electuary called Philonium Ro-
manum. This however was thrown up: for he
had already begun to vomit porraceous bile, which
afterwards became æruginous, and at laſt, near
his death, ſo black that it approached to the colour of ſoot. Within about ten hours from the
beginning

beginning of the pain, Valſalva was called. As
he found that the patient had an unpromiſing
look, his abdomen being tenſe and painful to the
touch, his pulſe being low, conſtricted as it were,
and not ſufficiently perceptible, his urine of a red
brown colour, and very turbid, and other bad
ſymptoms being preſent; and recalling to his
mind ſimilar caſes, he ſaw that much miſchief
had been done within a ſhort time, and prognoſti-
cated that the patient would die within twenty-
four hours. That the patient, however, might
not immediately underſtand this, he ordered him
to ſwallow freſh drawn oil of ſweet almonds, and
his belly to be anointed with oil of violets combined with camphor; and he deſired that two
older phyſicians ſhould be ſent for. When theſe
practitioners had conſidered the caſe, they were
of opinion that the patient was oppreſſed with convulſions; and therefore adviſed that blood ſhould
be drawn from the feet, and that a large cupping
glaſs ſhould be applied to the abdomen. Valſalva having modeſtly objected to this practice, the
advice of the two older phyſicians was adopted.
A vein was twice opened. From the firſt wound
nothing was diſcharged: from the ſecond a little
blood flowed out; but it immediately loſt its force,
and came out ſo ſlowly and feebly, that although
the vein was then inſtantly ſtopt, the pulſe could no
longer be felt. Slight delirium ſoon after ſuper-

vened: his eyes appeared somewhat convulsed; his respiration became difficult; and he died during the night, according to the prognosis of Valsalva.

Appearances on Dissection.

EXTERNALLY. When the abdomen was felt with the hand, it was obvious that some fluid was effused within it.

ABDOMEN. Extravasated blood, in a fluid state, to the quantity of a pound and a half, was found in the belly. A strong smell, though not excessively strong, issued from that cavity. The intestines were to a great extent, and especially those in the upper part, red in several places; and the ileum had already begun to grow livid. The peritoneum in some places, but more particularly where it invested the diaphragm, was marked with black spots. Where, however, it covered the external surface of the stomach, which organ internally was in a natural state, it was unequal from black tubercles rather than spots. These tubercles, although at first they appeared to be glands, were in fact nothing else but stagnant blood; or rather, they proceeded from an incipient gangrene.

THORAX. Some blood appeared effused into the bronchia.

CASE

CASE II. (XXXV. 12.)

A poor blind old woman, of a small and slender make, having been indisposed for three days, was brought into the hospital of Padua, affected, it was supposed, with inflammation of the thorax. Nothing certain could be learned from herself; for her strength was so much exhausted, and her pulse so very weak and small, that she died the very day she was brought in.

Appearances on Dissection.

ABDOMEN. The intestines were inflamed; as was also the liver. The internal surface of the uterus, probably from the same cause, was as red as if the woman had lately had the catamenia. The anterior and posterior internal surface of the cervix joined at an angle on the right side; and from that part a small membrane, though not very minute, was extended transversely to the posterior surface, adhering to it by its whole lower edge. The rest of it was loose and floating; and it lay in such a manner, that, contrary to the common appearance of the valvulæ cervicis, its concave surface was turned upwards, and not downwards. It was therefore probable that this membrane had not existed originally, but had been produced, perhaps, in consequence of difficult parturition; it was certain that the woman

had

had born children. The uterus inclined to the right fide.

THORAX. The lungs were found. The peri-cardium was every where connected to the heart, by a continued, but not very ftrong, cohefion; for the two could be eafily feparated by the fin-gers, without any laceration whatever. It did not adhere to the large veffels. On the furface, by which it had adhered to the heart, a white fpot appeared only in one part, and occupying a fmall fpace. The ventricles of the heart contain-tained blood of a black colour, as it was every where elfe in the body; but no polypous concre-tions. Concretions of that kind, however, were found in feveral veffels. Some of them were round and white; and fome of them were thick, firm, and even long, as thofe were which extend-ed from the right auricle to the internal jugular veins.

HEAD. In one of the eyes no morbid appear-ance occurred, except opacity of the cornea. In the other, although feveral parts were well form-ed, the chryftaline lens appeared fo much dimi-nifhed that very little of it remained. What did remain was white and opaque, yet internally moift. It, together with the iris, adhered to the cornea, where that coat was more opaque than natural, and fomewhat excavated into a fmall pit of an oval fhape, and was tinged of a dirty yellow

X x 2   colour.

colour. This difeafe, however, did not extend to the external furface of the cornea.

CASE III. (xxxv. 14.)

A YOUNG man, addicted to the ufe of wine and fpirituous liquors, who had lately laboured under intermittent fever, became affected with a pain in the belly, which was removed by the difcharge of wind downwards. The pain however returned within a few days; and he was brought into the hofpital of Bologna, on the fixth day after the return. The pain continued conftantly in the hypogaftric region. It was flight, and only from time to time became violent. At thefe times, the belly often fwelled confiderably in that part; and when it was examined with the hand, feveral hard globules were felt. All thefe fymptoms quickly difappeared, and again recurred at intervals. The ftomach was painful; and he vomited every thing received into it, not excepting the medicines, among which even was opium. As his belly could not be kept open, but by means of glyfters, it was determined that, while that circumftance fhould be held in view, both remedies and nourifhment, confifting of broth and emolient herbs, fhould be given by way of glyfter. This treatment, however, produced no alleviation of pain: befides, no ftool could be procured until an injection of lin-

feed

feed oil had been more than once exhibited. The belly was anointed with the fame oil, and with other fubftances, without any good effect. He felt the pain eafier, when fitting up in bed than when lying: therefore, he fat up even when he flept. He felt himfelf better too, and flept more eafily, with an empty ftomach, than when he had by chance retained any thing in it. This circumftance, and alfo the abfence of fome other fymptoms which generally mark the prefence of worms in the alimentary canal, rendered it probable that his pain did not proceed from that caufe, although he had, three days before this, vomited up a large lumbricus. At laft he began to retain fome nourifhment, and even no longer to vomit his dinner. His cheeks were red, which he himfelf attributed to a determination of blood to the face, to which he faid he was fubject. He was thirfty; and his whole abdomen was diftended. It was now the fifth day fince his admiffion into the hofpital; and in the morning of that day he faid he was fomewhat better; a circumftance which was confirmed by the appearance of his countenance, the tone of his voice, increafed ftrength of his body, as appeared from the manner he fat, and the ftate of his pulfe, which neither was then, nor hitherto had been bad. At that time he had no fever; nor had any feverifh fymptom, except perhaps on one occafion, been obferved during his refidence in the hofpital.

hofpital. Within two hours, however, notwithftanding thefe favourable appearances, he began
fuddenly to cry out from the feverity of pain, and
he continued to do fo for fifteen hours. In the
mean time, he had vomiting; and towards the
evening, he himfelf gave notice that his pulfe
could no longer be felt, which was really found to
be the cafe. At the end of fifteen hours from the
commencement of this attack, he faid that he
muft get out of bed to go to ftool. While at ftool
he fainted; and within half an hour died.

Appearances on Diffection.

When the body was wafhed the day after death,
a great quantity of fluid, like putrid blood, diluted with a very ftinking tobacco-coloured liquid
matter like fæces, was difcharged from the mouth.
The abdomen, in confequence of this, became
fomewhat flaccid about the hypogaftric region:
and although in the epigaftric region, which was
of a livid colour, and in the remaining parts, it
was ftill hard and diftended, it was lefs fo than before death.

ABDOMEN. Whenever the cavity of the belly
was opened, a great quantity of fluid, like that
difcharged from the mouth, burft out fo fuddenly,
that it was uncertain whether it had proceeded
from the cavity of the belly, into which it had
been previoufly effufed, or from a diftended inteftine, which might have been readily cut through

along

along with the peritoneum.   At any rate the cavity of the belly appeared full of that fluid.   All
the fmall inteftines were as black as coal.   The
fpleen alfo was fimilarly affected with gangrene,
at leaft in part.   The ftomach, however, and that
portion of the large inteftines extending from the
extremity of the ileum to the left hypochondrium,
were found, as far as could be judged from external examination ; for the intolerable ftench, which
was increafed by the fæces having paffed through
a wound made by carelefsnefs in one of the inteftines, prevented any more accurate examination.
Along with the fæces a lumbricus worm, of a moderate fize, was difcharged through the wounded
inteftine.

C A S E IV.   (XXXIV. 25.)

AN old man, aged feventy-four, of a lean habit of
body, addicted to drinking, had begun for a month
to walk in fuch a manner as to bear chiefly on his
left leg.   His fervants remarked this circumftance
more than himfelf ; and indeed he never fpoke on
the fubject, nor appeared to feel pain in any part.
Within eighteen days after, he was feized with a
wandering pain in the belly, unattended with fever ; which he himfelf, without any advice, expelled, by means of theriac.   But within the fpace
of twelve days after, he became affected at mid-

day

day with excruciating pain, exciting the fenfation, as he expreffed it, as if he were bitten by dogs, at the fuperior part of the iliac region, on the right fide. The pained part was fwelled, but was not difcoloured. When touched fuperficially, it felt foft; but when the hand was forced more deeply into it, a hardnefs was perceived. His pulfe was frequent, and quick in the contraction of the artery; but in other refpects good. His eyes appeared much funk; his tongue was parched; and he paffed a reftlefs night. On the fucceeding day his pulfe was more full, and was vibrating. The pain and fwelling extended to the middle of the belly, and at laft to the left fide alfo. Seven ounces of blood were drawn from the right arm. The blood contained no ferum, and exhibited a thick yellow cruft. He had naufea; but not to fuch a degree as to vomit his food. His belly was kept open without any trouble. After having paffed a very bad night, his pulfe, on the third day, was low; he had frequent bitter and four eructations; his voice was impaired, as if from convulfion; and his mind from time to time wandered, as the nonfenfe which he fpoke fully indicated. On the fourth day, his extremities were now and then convulfed; and his whole body remained rigid for a quarter of an hour. At that time no pulfe could be felt; but whenever the convulfion ceafed, the pulfe returned, and was

I   like

like that of a healthy perfon, except that it was
low, and when preffed upon by the fingers afforded no refiftance. Soon after, his refpiration
became very difficult; and although his tongue
was moift, and the delirium had ceafed, he vomited feculent matter; and in the evening of the
fame day he died convulfed.

Appearances on Diffection.

ABDOMEN. The left lobe of the liver appeared flabby, and throughout gangrenous. The ftomach and inteftines, efpecially the fmall ones, were
in fome places red, in fome livid, and in others
black. The beginning of the colon, at that part
where it was in contact with the mufcles that cover the foffa iliaca, and alfo thefe mufcles themfelves, were compleatly gangrenous. It was fo
ftrongly connected with thefe mufcles, that it
could not be feparated from them without laceration. A livid-coloured ferous fluid, mixed with
pus, which had been found in the cavity of the
abdomen, feemed to have been effufed from that
part of the inteftinal canal, as the inteftines contained a fimilar fluid.

CASE V. (XXXIV. 27.)

A WOMAN, who had had a fall on her back
about a year before, became affected with violent
deep-feated pain in her belly, attended with vo-

VOL. I.   Y y   miting.

miting. This having continued for fome days, fhe died.

Appearances on Diffection.

ABDOMEN. The ftomach was found wonderfully contracted. The caput cæcum coli was fo much diftended with femi-fluid yellow fæces, that it equalled the ufual fize of the ftomach. That inteftine was affected with inflammation; which had alfo begun to extend over the neighbouring vifcera.

C A S E VI. (XXXV. 10.)

A RUNNING footman, upwards of fixty years of age, of a fhort ftature, and fat habit of body, having been no longer able to act as a fervant, had begged for fome years, and had drank very freely whenever he could procure wine. On his return home one day, he complained of being unwell; but he took nothing by way of remedy, except bread and wine. Immediately after this, he complained of pain in his belly; which continued till mid-night, when he died.

Appearances on Diffection.

ABDOMEN. When the abdominal mufcles, which were flaccid, were cut through, and the cavity opened, a ftrong fmell was felt. A confiderable portion of the fmall inteftines defcended pretty low into the cavity of the pelvis, fo as to reach

reach the junction of the urinary bladder and rectum, and filled up the whole space therein contained. This, however, had been an original conformation, or at least not a recent one. That portion, and other parts of the small intestines, were in some places much contracted, and were there of a brown colour; but elsewhere they were red, the most minute vascular ramifications being as much distended with stagnant blood as if they had been injected with red wax. The large intestines were in the same state here and there, especially about the beginning of the colon. The edge of the liver was somewhat black. The spleen was larger than natural. The trunk of the aorta, within the belly, exhibited some small points of ossification. The vena cava was distended with much black fluid blood.

CASE VII. (xxxv. 16.)

A SLENDER woman aged forty years, of a short stature, and of a bilious temperament, having after a fit of passion become affected with pain in her side, was admitted into the hospital of Bologna. She had been a widow for three years; and had had no appearance of the menses for eight years. A spitting of blood, to which she had been subject from time to time, was imputed to this cause; though it appeared to the physician who attend-

Y y 2 ed

ed her, to proceed from the pharynx rather than from the lungs. Pain, which refembled that of the cutting of knives, was felt firft below the left breaft, and then, without leaving that fituation, it extended to below the right breaft; but there it was more flight, fo that fhe could lie upon that fide. The pain was aggravated when the part affected was touched; and it rendered her refpiration difficult. It had been ufhered in by febrile rigour, which, although the fever did not intermit, recurred every day. Her face was flufhed. She had great thirft; and fhe fuffered much from a cough, as it aggravated the pain. Her expectoration was frequently bloody; at other times white, thick, and frothy. She was fometimes affected with the fenfation of fomething rifing up to her throat. At laft fhe felt a pain about the umbilicus, as if fhe were torn by dogs. The belly was open. Blood was drawn from her foot; and other means deemed ufeful were employed. Within a few days after this, without any previous critical evacuation, all the fymptoms were fo much alleviated, that the phyfician pronounced her to be convalefcent. She arofe; but her ftrength having immediately failed, fhe was obliged to return immediately to bed, where fhe was found, with her limbs drawn up, and her body bent, in the pofition generally taken by thofe affected with the fenfation of cold, and without any pulfe. When

fhe

fhe was afked, whether fhe felt any pain in the thorax or abdomen, fhe replied in the negative. On the fame day, fhe began to pafs fetid blood by ftool. She afterwards grew delirious, and had fubfultus tendinum. In confequence of thefe circumftances, fhe became much weakened, fo that fhe could no longer fpeak ; and, on the fixteenth day from the beginning of the difeafe, fhe died.

Appearances on Diffection.

ABDOMEN. When the belly, which had become flat, was cut into, and the cavity laid open, a fmell, fuch as generally proceeds from gangrenous parts, with a combination of that kind of acid fmell felt where there are lumbrici, was emitted. Almoft all the fmall inteftines were of a red, approaching to a livid and black, colour ; and in them fome lumbrici were found. The fame livid colour appeared at the lower part of the flat furface of the fpleen, and penetrated pretty deeply into its fubftance. The pancreas had become thickened, and confifted of a kind of indurated globules. The liver was alfo fomewhat hardened. The gall-bladder was diftended with calculi, to the amount of an hundred and twenty, together with fome pale coloured bile. The largeft of thefe calculi, which were twenty in number, equalled a filbert. All the calculi had a pretty firm fmooth furface. They were of various fizes and fhapes; but they all, as is generally the cafe, approached

approached nearly to the form of a cube. One of
them being applied to a burning candle, at firſt
ſwelled and bubbled up, emitting no bad ſmell;
and then took fire. It preſerved the flame, melting
into drops, and from time to time, while burning,
ſparkled with a ſmart noiſe, and retained flame
to the very laſt; ſo that the flame proceeding
from it was more durable and more bright than
that of burning ſealing wax. Others were burnt
in the ſame manner, and exhibited the ſame phenomena. The uterus inclined ſo much to the right
ſide, in conſequence of the round ligament on that
ſide being very ſhort, that, looking at the middle
of the pelvis, no uterus was ſeen. A puſtule, of
the ſize of a lupin, filled with white purulent matter, projeſted from the uterus, at the inſertion of
the left Fallopian tube. When the puſtule was
opened, and the matter diſcharged, the ſubſtance
of the uterus which it hollowed out appeared black.
The Fallopian tubes contained matter which was
not white, but of a yellow fleſhy colour. The o-
varia were contraſted: within them there were a
few veſicles; and the coat of one of them was in
ſome meaſure cartilaginous.

THORAX. The lungs, on their anterior ſurface,
were conneſted to the pleura in a very few places by membranous ſubſtances. They were ſound;
excæpt the anterior part of the right lobe, the
ſubſtance of which was compaſt, but not in a ve-

ry

ry great degree. The pericardium contained no ferum. The heart was flabby. In the right ventricle, and at the orifices of all the veffels, fmall polypous concretions were feen. The pofterior part of the pharynx, oppofite to the epiglottis, was very much eroded : and the velum pendulum palati alfo at one part appeared black, rotten, and entirely perforated.

HEAD. When the head was feparated from the neck, a confiderable quantity of watery fluid flowed out from the great foramen of the os occipitis. On opening the head, a fimilar fluid was found under the pia mater, efpecially on the left fide. The lateral ventricles contained reddifh ferum. The choroid plexufes were rendered unequal by a number of hydatids, which readily burft on being touched. When the medullary fubftance was cut into, bloody points appeared : and when it was preffed, a greater quantity of blood than ufual was fqueezed out. The fame fmell of worms which was felt in the belly, was perceived alfo in the diffection of the brain, in the tongue, in the pharynx, and even in the very eyes.

CASE VIII. (xxxiv. 23.)

A MAN, aged fifty years, of a lean habit of body, and pale colour, having drank a great deal of wine along with fome fellow-drunkards, was

seized

feized with violent, but wandering pain, in the belly, attended with flatus, vomiting of bilious matter, and a quick pulfe. On the morning of the fucceeding day, as the pain was not only more fevere, but alfo fixed in one part, which was very fore to the touch, Valfalva being afraid of inflammation, ordered a vein to be opened. All remedies, however, were in vain : for he died about the beginning of the fourth day after the attack.

He had been two years before affected with acute fever, from which he had recovered without any evident crifis. Soon after this, he complained of much thirft ; he felt a very great weaknefs in the head and ftomach; and his ftrength became impaired. Along with thefe fymptoms he was fometimes diftreffed with confiderable oppreffion, which at night when he wifhed to go to fleep, was fucceeded by tremor of the whole body. It was imagined by fome that the man laboured under phthifis pulmonalis. But Valfalva thought that his complaints proceeded from a quantity of water in the cranium; and on that account, he prefcribed thofe medicines which are proper for dropfical patients.   •

Appearances on Diffection.

ABDOMEN. A large portion of the ileum was inflamed. All the other vifcera were found.

THORAX. A very large polypous concretion

3   appeared

appeared in the right ventricle of the heart, from whence it extended into the vena cava.

HEAD. A confiderable quantity of ferous fluid was found within the cranium; and the ventricles were full of the fame kind of fluid. The glands of the choroid plexufes were very large, and contained much ferum. The corpus callofum, and the other parts which joined the two hemifpheres of the brain, were flabby.

CASE IX. (XXXI. 25.)

A WOMAN died in confequence of dyfentery.
Appearances on Diſſection.

ABDOMEN. The inteftines were found in a ftate of inflammation. The right kidney was wanting; but the deficiency was fupplied by the left, which was twice the ordinary fize, and contained a double pelvis, and double ureter. Both ureters went to the right fide of the bladder.

CASE X. (XXIV. 16.)

AN old man, of a lean habit of body, was brought into the hofpital of Padua, on account of ftrangulated hernia. His pulfe was fmall and weak, but not intermitting; yet notwithftanding every means which could be employed, he died.

Appearances on Diffection.

ABDOMEN. The inteſtines were inflamed. The teſtis next the hernia was conſiderably leſs than the other. When cut into, its ſubſtance internally appeared of a brown red colour; but that of the other was natural. Between the ſound teſticle and the tunica vaginalis, there was a little quantity of watery fluid; and at one extremity of the ſame teſticle, a ſmall roundiſh body, like the remains of a ruptured hydatid, projected. The gall-bladder was placed tranſverſely; and although it was of the ordinary ſize, it was not received as uſual into a depreſſion of the liver; for after its fundus was ſeparated from the liver, (and this was done without the leaſt force being neceſſary) the part to which it had adhered could ſcarcely be diſtinguiſhed, being ſo ſmooth, that if any veſſel had connected them, it muſt have been ſo exceedingly minute as to eſcape the notice of the ſenſes. This was certainly an original conformation. It contained a little quantity of black, and ſomewhat viſcid bile, together with twenty calculi. Theſe were of a black colour; ſmall, but nearly equal to each other in ſize. They all conſiſted of ſeveral globules as it were; and being indented into each other, they were all in contact. When applied to a burning candle, they neither melted nor flamed, and ſcarcely even ſparkled. All the other abdominal viſcera were ſound:

THORAX.

THORAX. The left coronary artery appeared changed into a bony canal, to the extent of feveral fingers breadth, where it furrounded the bafis of the heart. A part of that long branch which is fent down from it, along the anterior furface of the heart, to the extent of three fingers breadth, was alfo offified. From this circumftance, the blood was tranfmitted on both fides, not through a membranous canal, nor through a veffel which had here and there points of offification, but through a continued bony tube, in a few places only fofter than in others; and thefe formed tranfverfe lines which might be compared to the joints of a fmall reed. When the heart was opened, and fome polypous concretions were removed, the tubercles of the valves of the aorta appeared much harder than ufual, and almoft offeous. No points of offification, however, were found in any of the valves, nor in the aorta near the heart. But the internal furface of that veffel, at fome diftance from the heart, both at the origin of the veffels going to the head and fuperior extremities, and alfo from that part quite to the divifion into the iliacs, was in many places unequal, from very hard bony fcales; feveral of which equalled in fize the nail of the thumb. The internal coat of the artery, however, which covered all thefe fcales, feemed injured only in one

place,

place, where there was an aperture, in which a thickiſh white matter was obſerved. Bony ſcales were alſo diſcovered at the origin of the ſubclavian and carotid arteries on the right ſide, and alſo in the iliacs, and in the ſplenic as far as its inſertion into the ſpleen. No offiſications, however, were obſerved in the arteries within the head, nor in the ſuperior extremities; thoſe in the latter were harder and ſomewhat wider than uſual. The blood which remained in the crural arteries was not fluid, nor yet was it polypous.

HEAD. Polypous concretions, of a pretty conſiderable thickneſs, were found in the lateral ſinuſes of the dura mater. In the right and left ventricles of the brain there was ſome ſerous fluid. On the choroid plexuſes of both theſe ventricles, hydatids appeared, ſome of which were pretty large.

C A S E XI. (xxxiv. 5.)

A MAN, aged forty years, of a ſanguineo-bilious temperament, who was ſometimes troubled with a ſlight hernia at the groin, became affeſted after eating artichokes with ileus. A ſlight tumor appeared at the groin: but he denied that he felt any pain there; and complained only of pain in the belly, which was very much indurated from the retention of the fæces. All the remedies that

were

were tried proved in vain; for on the feventh day fatal vomiting fupervened.

Appearances on Diffection.

ABDOMEN. The inteftines appeared turgid with air; and at that part near the cæcum where they are doubled, they were livid and black; and, together with the annexed portion of the mefentery, which appeared flefhy as it were, had fallen down to the extent of four fingers breadth into a hernial fac, with fo narrow an orifice, that, after they had been diftended with the matter contained in them, they could not have returned into the abdomen. This fac was in the right groin: and was formed by an elongated and dilated portion of the peritoneum; but not, as was formerly imagined, by that portion of it which accompanies the vas deferens and fpermatic veffels; for it lay on the anterior part of that procefs and of thofe veffels, which were very much diftended with blood. It was internally of a black colour, or rather of a black green, as if it had been ftained by vitriol, as a ftrangulated portion of the inteftine generally is. In the left groin there was another fac very much like that juft defcribed, except that the membrane of which it was formed was in every refpect natural.

THORAX. Polypous concretions of a yellow colour, along with coagulated blood, were found within the ventricles of the heart. That in the

3   right

right was larger than that in the left; but neither extended beyond the ventricles.

Case XII. (xxxiv. 9.)

A young man, a ploughman, who had been feven years before troubled with a hernia on the right fide of the fcrotum, having had the prolapfed inteftine replaced, and retained by means of a trufs, fuffered no inconvenience from it, until he had laid afide the ufe of the bandage, when, after having been affected with intermittent fever for two months, and having ftuffed himfelf with dumplins and other indigeftible preparations of unleavened flour, the inteftine again fell down as formerly. On that very day he began to have vomitings of bitter matter. To thefe, fingultus and pain of the fcrotum fupervened, on the fourth day of the difeafe. The fcrotum having been fomented with warm foap leys, the pain in it feemed alleviated. But as the vomiting and fingultus continued, and as he was befides affected with pains in his belly, together with thirft, he was brought on the fixth day into the hofpital of Bologna. The *emplaftrum de crufta panis* having been applied to the region of the ftomach, and an enema compofed of linfeed oil and of oil of violets, having been exhibited, the fingultus, and vomiting alfo, abated, though for a fhort time only.

On

On the feventh day the pain in the fcrotum was diminifhed. His pulfe was lefs frequent than it had been on the preceding day, but was weaker than it fhould have been in a young man ; his thirft continued ; and he paffed no fæces until the oily injection already mentioned had been exhited. On this day, injections, compofed of the Carminative Decoction, as it is called, to which clarified honey, together with two drachms of electuary, called Benedicta Laxativa, were added, having been adminiftered, the vomiting of bitter matter returned, and at the fame time a worm of the lumbricus kind was thrown up. The injection was not entirely paffed even after many hours had elapfed. On the eighth day another lumbricus was vomited. The abdomen, although it was tenfe, which it had been the day before, and although it refounded under the hand as if there were tympanites, was not painful to the touch, even though rudely handled ; except, indeed, in the epigaftrium, where he felt a kind of gnawing pain. When he was afked if he alfo felt heat in that part, he replied in the negative. His pulfe was nearly in the fame ftate as on the day before, except that it was much more frequent. His tongue was parched. His urine was of a deep colour. Under his eyes there was a livid mark ; and, independent of that, his face had an unfavourable appearance. He paffed a reftlefs night;

and

and on the ninth day was much in the fame ftate, though, indeed, his pulfe, and the appearance of his countenance, were rather worfe ; for his pulfe was fomewhat quicker, and the artery when preff-ed afforded little or no refiftance, and his face had nearly that appearance ftiled Hypocratic. Al-though on the preceding days he had had confi-derable anxiety, and fpoke in a defponding tone of voice, and frequently changed the pofture of his limbs ; on this day all thefe circumftances ap-peared more remarkably. Moreover, befides the conftant pain over his whole belly, he felt at in-tervals, in different parts of the abdomen, but efpecially in the epigaftrium, gnawing fenfations. He was afked if he had any throbbing pain in his belly, or if he felt throbbing in any part of his body ; and he anfwered in the negative. The pain which he felt in the fcrotum and contiguous part of the belly, was not, on thefe latter days, according to his eftimation, the chief pain. His fkin was dry and rough ; but not unufually warm. After having taken fome food, he felt better. He faid that he had been relieved the day before by the oily injeĉtion, which indeed he had men-tioned at that time. On this day he had another injeĉtion, confifting of broth in which coriander feeds had been boiled, and into which fugar had been put. When he paffed this, he vomited the food he had taken. Towards the evening he had

fome

fome fleep. After this, he complained of a kind
of throbbing fenfation in the epigaftrium, and
of fome fenfe of heat in the belly. In the mean
time, he vomited at intervals a yellow-coloured
matter, more liquid than that hitherto thrown up.
All thefe fymptoms having continued during the
whole night; on the morning, which was the tenth
day from the beginning of the difeafe, he died.

Appearances on Diſſection.

ABDOMEN. A great quantity of the fame kind
of matter as that vomited, was found extravafated
in the cavity of the abdomen; and the ftomach
and fmall inteftines, even as far as the hernia,
were greatly diftended with the fame kind of matter. Within all that tract of the inteftines, a fingle
lumbricus only, like thofe formerly vomited, was
difcovered. The large inteftines, which were empty, and of a white colour, were found. The ftomach
was alfo found. The duodenum, however, to the
extent of fix fingers breadth, had become fo livid
from inflammation, that it had a gangrenous fmell.
The jejunum, and by much the greateft part of
the ileum, were here and there affected with a
flighter degree of inflammation, as they were not
livid. The remaining part of the ileum, namely,
that, which lay neareft the colon, was rather gangrenous than inflamed. The hernial fac was of
the fhape of a pear; and was compofed of a
coat, which was not lefs thick and firm than

that of the pulmonary artery. It was covered not only by the fcrotum and dartos, but
alfo by the cremafter mufcle, and the membrane
on which that mufcle in common with the teftis
lies ; and alfo by the veffels belonging to the teftis.
The teftis lay under the fac ; and its veffels, which
adhered externally to the infide of the fac, paffed
into the belly near its orifice, but not through it.
That orifice was like a pretty thick ring, which had
been formed by the peritoneum and furrounding
tendon.　Befides the ileum, and part of the mefentery attached to it, it contained alfo the omentum, almoft no part of which had been feen covering the inteftines on the left fide, as it had been
drawn down on the right fide into the hernia.　It
not only extended to the bottom of the fac ; but
from thence forming itfelf into a round body,
(which, unlefs it had been cut into, could never have been known to have been compofed of
the compreffed fubftance of the omentum), it
returned upwards, and was connected to the
ftrangulated ileum, at no great diftance from the
orifice of the fac.　That portion of the omentum, thus prolapfed, was here and there connected to the fac, by certain interpofed red flabby fubftances, which could be eafily feparated both
from the fac and omentum ; and appeared to be
nothing elfe than membranous cells filled with
blood and ferum.　The ileum neither was connected

nected to the fac, nor did it extend to its fundus;
but being reflected a little below the orifice, it returned into the belly by the fame paffage by
which it had come out: fo that not more than
four or five fingers breadth of the inteftine was
ftrangulated. All that portion being affected with
gangrene, was of a black colour; and that part
conftricted by the orifice of the fac was much more
black and gangrenous; as was alfo the ring forming the orifice of the contiguous part of the ileum,
lying above it, which was fo rotten as to have been
incapable of bearing the weight of the fluid that
had diftended it: for the fluid had efcaped through
a pretty large opening into the cavity of the belly.
The edge of the liver was black; and its concave
furface, together with the gall-bladder, which was
fmall, were of a blackifh colour. All the contents
of the belly were not a little warm, although the
body was not opened till thirteen hours after
death.

THORAX. A foft yellow polypous concretion.
was found in the right ventricle of the heart, from
whence it fent out white-coloured branches even
as far as the jugular veins.

CASE XIII. (XXXIV. 14.)

A WOMAN, above fifty years of age, who had,
for thirty-two years been affected with two herniæ,

both

both on the left fide, the one at the umbilicus, and the other at the pubis, having accidentally fallen, received a contufion about the top of one of the fcapulæ and the extremity of the fhoulder bone. Although fhe readily recovered from this bruife, fhe began in a few days after the fall to be coftive, and foon after to vomit a yellow fluid matter of the fame fmell, as fæces. The vomiting, although it occurred at other times, took place chiefly about two or three hours after having taken food. Her pulfe was neither frequent, nor did it afford very little refiftance when the artery was preffed; but it was exceedingly fmall, particularly after vomiting; and it became fmaller every day. As glyfters produced no effect, mercury, to the extent of two drachms, was twice given. The firft dofe proved of no ufe; but by means of the fecond fhe had three ftools, the two firft of which confifted of indurated fæces, and the latter of fluid feculent matter. The medicine feemed to have no bad effect. Neverthelefs, about twelve hours after having taken the fecond dofe of mercury, that is, within four or five days after the vomiting had begun, having half an hour before had a return of the vomiting, fhe died. During the whole courfe of the difeafe, fhe had neither been affected with obvious fever, nor with convulfions, nor had fhe complained much of pain in the belly.

Appearançes

Appearances on Diffection.

ABDOMEN. When the belly was opened, a strong smell was perceived. The jejunum and contiguous part of the ileum were quite distended with the same matter which had been vomited. The remaining part of the ileum, and the large intestines, were contracted. The jejunum, in some places, was marked longitudinally with streaks of a bright red colour; in other places it was of a red brown colour, as the ileum was almost every where. That latter intestine, not far from the jejunum, to the extent of three or four inches, formed into an arch, together with the annexed mesentery, had fallen down into the sac of the lower hernia. Although it neither adhered to the sac, nor was compressed by its orifice, which formed a kind of ring, it had become gangrenous. It was of a black bloody colour; and bloody serum distilled from its surface. The upper hernia, when looked at externally, seemed divided into two little eminences. Internally, it was found to consist of one sac only, formed by the peritoneum, into which nothing more than a portion of the omentum had entered. The liver was somewhat harder than usual. The spleen was flabby, and externally appeared in some places livid. The ligaments of the uterus were black ; that organ itself was very small, and had thin parietes. On cutting into it, the substance of the parietes in the middle was so livid, that it

appeared

appeared approaching to the ftate of gangrene.
As the uterus was fituated a little lower than ufual, the ftate of the vagina was examined, in order
to afcertain how far the uterus had fallen down.
When the labia were feparated, and the orificium
vaginæ brought into view, a body was feen within it, which at firft might have been taken for the
os tincæ. But as the uterus had not appeared
placed fo low as to reach that part, even although
it had been very large, the vagina was immediately taken out of the body, that the difeafed appearance might be more accurately examined. It was
then found, that the glandular body of the urethra, called the Proftate, had become very thick,
and had drawn down the vagina, which was flaccid and deftitute of rugæ, fo low, that its extremity, where perforated by the orificium urethræ,
might have been readily miftaken, by an unfkill-
ful perfon, for the os tincæ.

C A S E XIV. (XXXIV. 15.)

A WOMAN, aged thirty-nine years, of a pretty
good habit of body, having not a bad colour, and
much lefs a jaundiced appearance, the mother of
feveral living children, was fubject to a fmall fe-
moral hernia, which fhe was accuftomed to re-
place herfelf whenever it became troublefome.
She had nurfed a child for fix months, when fhe

had

had a fit of the hernia, from which she could not
relieve herself as formerly. After having for several days attempted to reduce the prolapfed inteftine, she became affected with fever, vomiting,
and the other fymptoms which attend ftrangulated hernia, except that she could always pafs fome
little by ftool. She was brought into the hofpital
of Padua, though too late, as she looked like one
juft at the point of death. Neverthelefs she dragged out her exiftence for feveral days, on the
laft of which she even feemed to be better, and to
be relieved, by the glyfters which were exhibited.
On that day, however, she died.

Appearances on Diffeƈtion.

ABDOMEN. The hernial fac, which was thick,
and made up of many laminæ, eafily feparable
from each other, being brought into view, was
found to be entirely unconneƈted with the round
ligament of the uterus, but to be attached to the
crural veffels, on the infide of which it lay. Its
orifice was not narrow; and therefore the com-
preffion of the prolapfed inteftine proceeded from
the lower edge of the external oblique mufcle,
called Poupart's ligament, that lay over it. Under
this ligament fome part of the colon was prolapfed. The inteftine, however, remained fufficiently open, except at the orifice of the fac, where it
was rendered impervious. At that part it was in
contaƈt with the fac, and was black and putrid.
· The

The contiguous portion of the inteſtine without the ſac was green. The parietes of the abdomen, internally, were alſo of a green colour, and had a ſtrong ſmell in moſt places. The gall-bladder was ſomewhat larger than natural; and contained, along with ſome bile of a black colour, ſixteen calculi. All theſe were ſmall, though not very much ſo; and were nearly equal in ſize to each other. Externally they were yellow, and they had ſeveral ſmooth ſurfaces. One of theſe calculi, in its wet ſtate, being applied to a lighted candle, burnt with ſparkling, and melted, but did not preſerve the flame.

THORAX. The whole left lobe of the lungs was connected to the pleura on one ſide, and to the mediaſtinum on the other. The thyroid gland was larger than natural. At the orifice of the pulmonary artery, inſtead of three valves, four were ſeen. They were all of the natural appearance, except one which was larger in every dimenſion.

HEAD. A great many bloody points appeared in the medullary ſubſtance of the brain, in conſequence of a large quantity of blood being accumulated within the cranium, as was evident from the diſtended ſtate of both venæ cavæ, and of the veins running into them, eſpecially the vena azygos.

2 CASE

C A S E XV. (xxxiv. 18.)

A PORTER, fo much worn out by conftant working, that although he was only fifty years of age
he appeared much older, had a hernia in the
right groin, of the fize of one's thumb, which
fometimes feemed to difappear. Without any
previous caufe, except perhaps that a ftorm of
fnow had fuddenly occurred after mild weather,
he became affected with a wandering, but acute
pain, in the belly; as if, to ufe his own expreffion, he were torn by dogs. By the application
of fome kind of ointment to the abdomen, the
pain feemed to remit; but it foon began a-frefh,
and was never afterwards alleviated. On this account he was brought, on the fixth day of the difeafe, into the hofpital of Bologna. At that time
his fkin was not hot, nor was his pulfe very frequent; but it was fmall, and when the artery was
preffed with the fingers, it gave little refiftance,
and its pulfations were found to be of unequal
force. His whole abdomen was as tenfe as a
drum; more efpecially below the right hypochondrium, where fome cells of the colon could, it was
thought, be felt by the hand, a pretty hard hernia being there formed, although he denied that
the principal feat of the pain was in that part.
He vomited his food. For four days he had had

VOL. I.   3 B   ·   no

no ſtool; and had not even been able to expel wind from the inteſtines, though he made many efforts for that purpoſe. Freſh drawn oil of almonds was given him, and a glyſter conſiſting of ten ounces of linſeed-oil was exhibited. The latter was paſſed in the ſame ſtate as it had been injected; the former was thrown up, and he complained that he had been much diſtreſſed thereby. When aſked what kind of taſte he felt in his mouth, he replied, that of poiſon. He had great thirſt; and the vomiting continued. On each of the two following days, which were the ſeventh and eighth, a glyſter was exhibited; that on the former day was compoſed of Laxativa Benedicta; and that on the latter, of milk and the yolk of an egg: but they produced no more effect than the former. No fæces being paſſed; the other ſymptoms having continued; the pulſe, although after the ſixth day it had been no longer irregular, having become weaker, and ſmaller, ſo that it could ſcarcely be felt; the ſkin being ſhrivelled, the body cold, and he being no longer able to raiſe his eye-lids, and ſcarcely to ſpeak, except to aſk for wine, he gradually ſunk; and on the evening of the ninth day, died in a placid manner.

Appearances on Diſſection.

EXTERNALLY. The body had a filthy appearance;

ance; and the skin, which was rigid, was not free from a scabby eruption.

ABDOMEN. When the belly was opened, a gangrenous smell issued forth. The omentum, as far as it extended into the hernia, was of a red colour from inflammation, except in a few broad transverse lines. The spleen in some part was of a morbid livid colour, which penetrated its substance, though to no great depth. The stomach extended much more to the right side than usual; being completely distended with a yellow matter resembling nothing more than fluid fæces; and the small intestines, as far as the hernia, were also distended with the same fluid. Whatever commonly passes from the ileum to the large intestines had remained in it, and that was in very considerable quantity. All the large intestines were contracted, and were of a white colour; by which it was evident that there had been no passage through the prolapsed portion of the ileum; although the tube of the intestine itself had not entered the hernial sac, but having passed by the side of it, only sent into it a portion of its paries stretched into the form of a semi-oval cavity. The one axis of this cavity, where it began gradually from the intestine, measured about three fingers breadth, and was in the longitudinal direction of the intestine; and the other, extending between the anterior surface, at the distance of less than an inch

from

from the infertion of the mefentery, and the inferior furface, was much fhorter. From its beginning. this cavity was contracted more and more,
as its femi-oval figure required, until at the middle, where it meafured one inch in depth. This
part, whether it be called a cavity or a cell of the
inteftine, was the only portion of the inteftinal canal interrupted by the hernia, into which alfo the
extremity of the omentum was prolapfed. Neither of thefe parts could be drawn back from the
hernial fac, becaufe they were not only fhut in
by the tendinous orifice of the fac, but were alfo
tied down to the fac itfelf, by bands, which although not very ftrong, were very numerous.
The fac on its internal furface was fmooth, except at the part connected with the hernia, where
it was fomewhat rough. It was formed by the
peritoneum, ftretched outwards near the outfide
of the fpermatic veffels. On each fide of the hernia there was a fwelled inguinal gland; one of
which was nearer the fac than the other, and
feemed compofed in part of a white fubftance.
That part of the inteftine next the fac, and ftill
more that in contact with it, was of a black red colour. The inteftine above the fac, (for below, as
mentioned, it was white,) was to a confiderable
extent of a red colour, approaching to livid; and
from that, as far as the ftomach, it was evidently
red from the fanguiferous veffels being in many
places

places much diftended. The mefentery was of the fame colour.

THORAX. The lungs adhered everywhere to the pleura, (excepting the right lobe at the anterior furface) but more efpecially at the fides and at the back, where that membrane was thicker than natural. The right lobe, at the upper part, where it adhered very ftrongly to the pleura, was exceedingly hard, as if from an old difeafe.; and its lower part was alfo not a little firmer than natural. The lungs were almoft everywhere full of fluid. In the pericardium there was no ferum. The heart was flabby; and at each of its orifices, as well as in the right ventricle and left auricle, it contained fmall tender polypous concretions. The fmalleft of thefe was that in the left auricle; and the largeft that in the pulmonary artery, extending from thence into its ramifications.

CASE XVI. (XLIII. 27.)

A MAN affected with all the fymptoms of ftrangulated hernia, was brought into the hofpital of Bologna. The difeafe having been too far advanced to yield to any treatment, he died.

Appearances on Diffection.

ABDOMEN. The hernial fac, included within the cremafter mufcle and tunica vaginalis, lay behind the fpermatic veffels and tefticle. Within
the

the fac there was a double portion of the ileum connected to it; but fo flightly, that it could be readily feparated by the fingers. This prolapfed portion could not be returned into the abdomen, on account of the ftraitnefs of the ring through which it had paffed, and the quantity of matter with which it was diftended. The ring and the inteftine within the fac, as alfo the contiguous portion within the abdomen, to the extent of half an ell, were of a black colour. The reft of the inteftines were not turgid, although before death the belly had been fomewhat fwelled. Within the tunica vaginalis of the teftis, on the fide op-pofite to that containing the hernia, there was about a third of a table-fpoonful of watery fluid. From the tunica albuginea, which in other re-fpects was, as well as the tefticle itfelf, in a found ftate, a fmall roundifh body projected. This was of the fame colour, and feemed compofed of the fame fubftance, with the coat itfelf.　　　-

HEAD. Serous fluid was found effufed within the cranium. The veffels of the dura and pia ma-ter were much diftended with blood.

CASE XVII. (XXXIV. 21.)

A MAN, aged fifty years, of a lean habit of bo-dy, after having undergone much fatigue in hunt-
ing,

ing, began to complain of the fenfation of great
heat in the throat and breaft. This fenfation having ceafed in thefe parts, was transferred to the
belly and loins; and being there accompanied
with an acute pricking pain, rendered the patient
fo uneafy, that he could not bear the affected parts
to be touched. During the firft days of the difeafe
he had frequent rigors. Within five or fix days
before death, fymptoms of volvulus fupervened,
attended with vomiting of the fæces; by which
his ftrength being gradually impaired, he died on
the thirtieth day after having been confined to
bed.

Appearances on Diffection.

ABDOMEN. The belly was found almoft completely filled with fanious fluid, which had connected the omentum and inteftines to one another.
The inteftines were much inflamed; as were alfo
the liver and fpleen; which latter was more deeply inflamed than the liver. The left kidney contained under its proper membrane fome extravafated blood; this did not however extend over
the whole furface of that membrane. Numerous
fmall abfceffes and ulcers appeared over the omentum and edge of the mefentery, more efpecially at
that part where it was connected with the colon.

CASE

C a s e XVIII. (lix. 15.)

A man, who appeared to be under fifty years
of age, of a well-fhaped body, fomewhat fat, of a
healthy complexion, a little inclining to brown,
and whofe hair and beard were black, having
been cured in the hofpital of Bologna of melancholic delirium with which he had been affected,
took, on the day before he was to have left the
hofpital, half a drachm of the extract of black hel-
lebore. This extract was prepared by pounding
the recent roots with pure water; and was often
exhibited to patients, without any bad effects, in
the quantity of a fcruple, and fometimes to thofe
whofe belly was not eafily opened, even in the
quantity of more than half a drachm. When this
medicine was prefcribed, the perfon taking it was
ordered to drink, during its operation, cow-milk
whey; but this patient had not done fo: he had taken nothing elfe than the extract. In confequence
of this dofe he had feveral ftools. In the beginning
of the night, that is, about feven or eight hours
after he had fwallowed the extract, when no mif-
chief was expected, he became affected with vomiting and pains in the belly. Thefe feemed alleviated foon after having taken fome warm broth,
that is, about the fecond hour of the night; but
at the fifth hour the fame fymptoms recurred.

3   Though

Though he vomited no more than two or three table-fpoonsful of a green blackifh kind of matter, they again feemed to have abated fo much that he went to bed before the fixth hour, and at that time appeared to be quiet and eafy, at leaft he made no noife that indicated pain, fo far as the patients in contiguous beds could obferve. Neverthelefs, at the eighth hour, a kind of found iffuing from his mouth having been heard, the attendants ran to his bed, where they found him already dead.

Appearances on Diffection.

EXTERNALLY. The limbs were not rigid nor contracted.

ABDOMEN. The ftomach and the inteftines externally appeared in different places inflamed. The ileum, in fome places, was of the natural width; and in fome narrower, and in others wider, than ufual. Where it was narrower than ordinary, the coats were very thin, and not of a red colour; in other parts they were marked with red ftreaks. The ftomach and the inteftines were firft wafhed out by water being paffed through them, and then opened. The ftomach, together with a fmall portion of the annexed gullet, feemed to be inflamed on the left fide only. The inteftines, here and there, were affected with inflammation, which however was lefs confiderable in the fmall than in the large guts, except in the rectum; and

VOL. I.   3 C   in

in it certain fpaces were as much inflamed as thofe
in the ftomach. No violent degree of inflammation was feen in any part of this fubject. The
fpleen was larger than ufual, and was of a rofy
colour. At the part in contact with the ftomach,
it was fo flabby, that, when cut into, its internal
parts feemed to flow out like a fluid. Nothing
extraordinary was obferved in the liver, except
that the bile appeared through the coats of the
gall-bladder to be of a light green colour.

THORAX. The lungs were found, and totally
unconnected with the pleura. Something like a
flender polypous concretion was found in the
heart. The large veffels contained little blood.

HEAD. When the upper part of the cranium
was removed, a fmall quantity of bloody ferous
fluid was difcharged. The finufes of the dura mater,
and the large veffels of the pia mater, contained
little blood. The brain, which was furprifing in
one who had been affected with melancholic delirium, was fo exceedingly flabby, that, when completely taken out of the cranium, and placed upon a table, the hemifpheres falling outwardly at
each fide, tore the poilerior part of the corpus ca-
lofum, although it was no more than fix days after the death of the patient. Notwithftanding
that laceration, the fafciculus, which appears longitudinally through the middle of the corpus ca-
lofum, was feen in a natural ftate, on the upper

3

part,

part, where it was entire. When the brain was
cut into, it was found that the veffels, both in the
medullary fubftance, and in the choroid plexufes,
were not deftitute of blood ; but a great degree of
flaccidity was obferved in every part, as well of
the brain as of the cerebellum and medulla oblongata. The pineal gland, which was rather larger
and of a more globular form than ufual, was alfo
flabby. Notwithftanding the exceffive laxity of
all the other parts, the arch which joins the right
and left fide of the third ventricle was not in the
fmalleft degree lacerated.

CAUSES of Enteritis.

Predisponent Cause. This does not feem to
depend fo much on the general ftate of the fyftem
which predifpofes to inflammatory complaints, as
on fome peculiarity, either natural or acquired,
of the inteftinal canal *. Perfons advanced in
years are more liable to enteritis than others.

3 C 2   Exciting

---

* Morgagni relates the following cafe, which marks very
clearly how fufceptible of inflammation the inteftines are, in particular ftates of the fyftem. " An unmarried woman affected with
cholic pains, unattended with any fymptoms of fever, having been
much relieved by the operation of a glyfter, which had brought
off fome bilious matter, and having become every day better,
was no longer vifited by her attending phyfician. Her fervants,
however,

Exciting Causes. Befides all the circumftances enumerated as exciting caufes of gaftritis; the
ftrangulation or conftriction of any portion of the
inteftines, in confequence either of hernia or in-
trofufception, (that is, the paffing of one part of
a gut within another) and dyfentery, induce enteritis.

————

REMARKS on the Cases of Enteritis.

The firft cafe affords a good illuftration of the
rapidity with which inflammation of the inteftines
deftroys life †. It fhows, too, how foon the favourable opportunity for blood-letting paffes away
in that difeafe.

The

however, having given a fuppofitory compofed of honey, inftead
of an enema, which they were accuftomed to exhibit every fecond
night, fhe was immediately feized with an excruciating pain in
the anus; and in the morning her pulfe could not be felt. Along
with the pain, there was fo great a degree of conftriction of the
anus, that a glyfter could not poffibly be adminiftered. When attempts were made, by means of emollient and anodyne remedies,
to remove the pain and conftriction, fuddenly as great a relaxation of the affected parts as is met with in dead bodies took place;
and about noon of that day fhe died."

† Morgagni in his remarks upon this cafe mentions, that a
monk at Bologna, who, although old, was very ftrong, died within twelve hours from the firft attack of enteritis, notwithftanding
every means that could be employed.

The third cafe proves how infidioufly enteritis often invades: for although MORGAGNI imagined that a remiffion had taken place, yet, as there had been no fymptoms of fever previous to the attack that proved fatal, it is perhaps more probable that the difeafe till then only threatened.

In the feventh cafe, the progrefs of gangrene was very clearly marked *. The worms obferved in this and in the twelfth cafe, are to be regarded, not as exciting caufes of the difeafe, but as an accidental circumftance.

The ninth cafe is an example of dyfentery proving the exciting caufe of enteritis.

The tenth, and following cafes, as far the feventeenth, are all inftances of the difeafe having been produced by hernia †.

In the feventeenth cafe, not only was the mefentery affected, but alfo in fome degree the kidney.

The laft cafe is a ftriking example of the pernicious effects of draftic purges in particular ftates of the fyftem.

In nine of the cafes, viz. the fecond, fourth,

fixth,

---

* The fmell that arifes from the prefence of worms, fo diftinctly perceived in the different parts of the fubject of this cafe, is a curious circumftance.

† No remarks upon hernia can be introduced in this part of the work, as they belong to local difeafes.

fixth, eighth, tenth, thirteenth, fifteenth, feventeenth, and eighteenth, the patients were above
fifty years of age.

SECT. VII.   *NEPHRITIS;* or, Inflammation of the Kidneys.

INFLAMMATION of one or both kidneys, is
known by the ordinary fymptoms of inflammatory fever, being attended by dull or acute pain in
the region of the kidneys, fometimes fhooting a-
long the courfe of the ureters, commonly accompanied with vomiting, and the frequent difcharge
of fmall quantities of deep red-coloured urine*, and
fometimes alfo with coftivenefs.

In fome cafes, a fenfation of numbnefs is felt in
the leg on the affected fide, and the correfponding tefticle is drawn up; but thefe fymptoms
generally indicate calculus in the kidney or ureters.

The difeafe is diftinguifhed from rheumatic affection of the lumbar region, by the patient being
able to bend his back without fuffering excruciating pain.

Nephritis differs from calculus in the kidney or
                                              ureter

* The urine in fome cafes too is quite limpid.

ureter (which has been ftyled Nephralgia) by the
fymptoms of fever accompanying, or immediately
following the attack of pain, and continuing without any remarkable intermiffion. Whereas, in
nephralgia they do not occur until a confiderable
time after violent pain has been felt, and they
frequently difappear entirely. In the latter cafe,
too, the numbnefs of the thigh, and retraction
of the tefticle on the affected fide, always take
place,

In nephritis the pain is often extended over
part of the belly, and imitates that of enteritis.
But, as the bowels are always either open, or eafily rendered fo by glyfters, the line of diftinction
between the two difeafes is readily drawn. Sometimes indeed both are combined.

The terminations of nephritis are the fame as
thofe of inflammation of the other vifcera within
the abdomen.

Refolution may be expected, if the pain abate
and the feverifh fymptoms diminifh. The crifes
are, profufe univerfal fweat, copious difcharge of
thick loaded urine, and hæmorrhagy from the feat
of the piles.

Suppuration is to be feared if the pain continue
violent, and become throbbing, and if then frequent rigors fupervene. When purulent matter
is paffed off with the urine, no doubt can remain.

Gangrene is a very uncommon event in this difeafe.

cafe. It is marked by the general fymptoms that characterife it in other cafes.

Sometimes the patient is carried off merely by the inflammation being communicated to other vifcera. In fuch cafes, the urine is fuppreffed, and fymptoms of enteritis take place.

The inflammation is commonly feated in the fubftance of the kidney, and feldom in its proper capfule *. In fome cafes, a fingle abfcefs of a prodigious fize is formed in it; in others, there are many abfceffes.

CASE of Nephritis.

(XXXVI. 20.)

An unmarried woman was affected with exceffive vomiting, attended with fever. The vomiting ceafed, but the fever remained; and a violent pain under the falfe ribs having fupervened, fhe died within two days.

Appearances on Diffection.

Abdomen. Some very limpid ferous fluid was found within the cavity of the belly. The ftomach and inteftines were much diftended with
air.

* Vide Dr. Baillie's Morbid Anatomy, pag. 178.

air. Each kidney was enlarged to more than three times its natural fize. The left kidney contained between its proper invefting membrane and cortical fubftance, efpecially in that part which was towards the fpleen, a fmall quantity of fanious matter.

THORAX. In the cavity of the cheft there was a little watery fluid. The lungs were found, except that they were marked with a very few black fpots. The pericardium was full of ferous fluid. From the ventricles of the heart fluid blood was difcharged ; the right ventricle, however contained an incipient polypous concretion.

CAUSES of NEPHRITIS.

PREDISPONENT CAUSE. Many circumftances are mentioned as predifpofing to nephritis : fuch as the ftructure of the kidney having become fo altered, as to be very fufceptible of inflammation ; hence, old perfons are more liable to the difeafe than others : the habit of conftantly lying on the back : fedentary life : the frequent or immoderate ufe of fermented liquors, &c. Like other vifcera, too, after they have been once inflamed, the kidneys are very liable to be again affected in the fame manner.

EXCITING

EXCITING CAUSES. All the general exciting causes of inflammations may induce neph___; but it is most commonly occasioned by some circumstance immediately affecting one or both kidneys. Thus, violent exercise on horseback or in a carriage, external injuries, over-exertion in some particular exercises or occupations, the use of large doses of diuretic medicines, &c. are the most frequent exciting causes of the disease.

THE case of nephritis affords no room for remark.

END OF VOLUME FIRST.

www.ingramcontent.com/pod-product-compliance
Lightning Source LLC
Chambersburg PA
CBHW021348210326
41599CB00011B/797